WATER CHEMISTRY

PRACTICES IN THERMAL POWER STATIONS

MJP
PUBLISHERS

WATER CHEMISTRY

PRACTICES IN THERMAL POWER STATIONS

Dr. K.K. SIVAKUMAR

Associate Professor and Head of the Department
Department of Chemistry
AMET Deemed to be University, Chennai.

MJP PUBLISHERS

Chennai New Delhi

MJP PUBLISHERS

This edition has been published in india by arrangement with Carson Books, UK

ISBN 9789355276773 **MJP Publishers**

No. 44, Nallathambi Street,
Triplicane, Chennai 600 005

MJP 1624

Publisher : C. Janarthanan

FOREWORD

Thermal power stations (TPS) meet the bulk demand of global electric power. In meeting this target, importance of contribution of good water chemistry is well recognised. During his more than 40 years of involvement under various capacities in keeping a satisfactory water chemistry regime for operating TPS, the author has experienced many ups & downs to feel the need of broad based education on water chemistry selective to thermal power industry.

Presently available books in the area deal with industrial water chemistry in general. The practical requirement of operating TPS is felt to be much ahead. With ever increasing demand of power energy, there has been continued mushroom growth of thermal power stations leading to opening of even new courses for engineering colleges as 'Thermal Power Engineering'. This book is therefore likely to fill the gap, and is dedicated to preservation & enhancement of our thermal energy.

The book, though can be originally seen as mostly compilation of the author's various lectures as delivered at different platforms, towards training programmes of TPS personnel of the country with respect to water chemistry aspects, but has been further augmented with relevant materials on equipments & the complex nature of chemistry of water & steam at high pressure & temperature, to avoid premature failure of the equipments ultimately leading to generation loss. Wherever found necessary, the author has tried to give clear concepts to its readers by providing theoretical backgrounds.

Hence, the author feels that this book may create & sustain the interest of the reader in view of the book being more practical than theoretical, giving special attention to describe trouble shooting parts in operating different equipments/ plants with suitable remedial measures and also with due interpretations of the operational and test data to overcome the operational problems of water treatment plant, boiler & turbine, etc. even with some of their basic design deficiencies.

The book is written with six different chapters with the following heads:

(a) Water treatment chemistry
(b) Boiler feed – steam cycle chemistry
(c) Cooling water chemistry
(d) Chemistry of water Pollution
(e) Case studies of few TPS
(f) Water testing methods

The book is therefore dedicated to thermal power industries of the country, wishing their performance at best, with a very satisfactory water chemistry results.

CONTENTS

A. Water Treatment Chemistry

B. Boiler Feed – Steam Cycle Chemistry

Chapter - A

WATER TREATMENT CHEMISTRY

This chapter deals with impurities present in water, problems associated with the use of untreated or inadequately treated water in boilers and related equipments, and the methods used to remove them. Brief description of equipments commonly used in water treatment has also been given along with instructions for routine operation and maintenance of these equipments.

An Overview of Water and Waste Water Management Practices in Thermal Power Plants

Thermal power plants generate electricity. Coal, fuel oil or natural gas is burnt to heat water in the boiler and generate steam which runs the turbine, which in turn is connected to the pulsator and generates electricity.

Exhaust steam is condensed in a water cooled condenser and is collected in the hot well. Demineralised water is added to the hot well to make-up the loss of water due to blowdown from the boiler and other system losses. Condensate is returned to the boiler via thermal deaerator by a condensate extraction pump.

Based on duty conditions, thermal power plants are classified as under.

- **Base load power plants:** run nearly continually to provide that component of system load that doesn't vary during a day or week. Base load plants can be highly optimised for low fuel cost, but may not start or stop quickly during changes in system load. Examples of base-load plants are modern coal-fired plants with a predictable supply of water.
- **Peaking power plants:** meet the daily peak load, which may only be for one or two hours each day. While their incremental operating cost is always higher than base load plants, they are required to ensure security of the system during load peaks. Peaking plants include simple cycle gas turbines and sometimes reciprocating internal combustion engines, which can be started up rapidly when system peaks are predicted.

- **Load following power plants:** can economically follow the variations in the daily and weekly load, at lower cost than peaking plants and with more flexibility than base load plants.

A simplified version of the steam turbine cycle is shown in figure below.

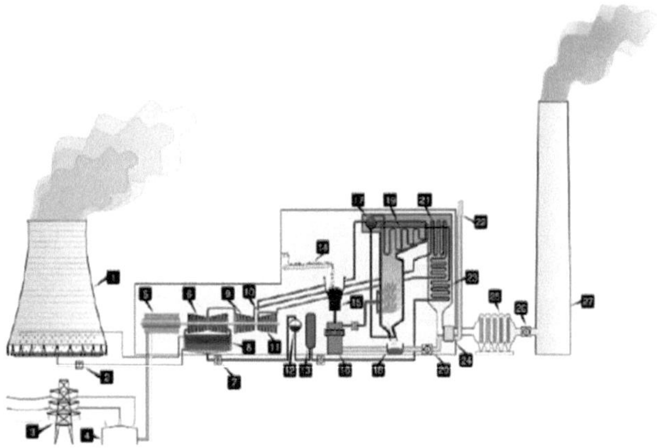

Figure 1

1. Cooling tower
2. Cooling water pump
3. Transmission line (3-phase)
4. Unit transformer (3-phase)
5. Electric generator (3-phase)
6. Low pressure turbine
7. Boiler feed pump
8. Condenser
9. Intermediate pressure turbine
10. Steam governor valve
11. High pressure turbine
12. Deaerator
13. Feed heater
14. Coal conveyor
15. Coal hopper
16. Pulverised fuel mill
17. Boiler drum
18. Ash hopper
19. Superheater
20. Forced draught fan
21. Reheater
22. Air intake
23. Economiser
24. Air preheater
25. Precipitator
26. Induced draught fan
27. Chimney stack

Description of a typical coal-fired thermal power plant

1. Coal is conveyed (14) from an external stack and ground to a very fine powder by large metal spheres in the pulverised fuel mill (16).
2. It is mixed with preheated air (24) driven by the forced draught fan (20).
3. Hot air fuel mixture is forced at high pressure into the boiler where it rapidly ignites
4. After a high purity flows vertically up the tube-lined walls of the boiler where it turns into steam, and is passed to the boiler drum, where steam is separated from any remaining water
5. The steam passes through a manifold in the roof of the drum into the pendant super heater (19) where its temperature and pressure increase rapidly to around 200 bar and 57°C, sufficient to make the tube walls glow a dull red.
6. The steam is piped to the high-pressure turbine (11), the first of a three-stage turbine process.
7. A steam governor valve (10) allows for both manual control of the turbine and automatic set point
8. The pressure and temperature of exhaust steam from high-pressure turbine is reduced before returning to the boiler reheater (21).
9. Preheated steam is then passed to the intermediate pressure turbine (9), and from there passed directly to the low pressure turbine set (6)
10. The exiting steam, now a little above its boiling point, is brought into thermal contact with cold water (pumped in from the cooling tower) in the condenser (8), where it condenses rapidly back into water, creating near vacuum-like conditions inside the condenser chest.
11. The condensed water is then passed by a feed pump (7) through a deaerator (12), and prewarmed, first in a feed heater (13) powered by steam drawn from the high pressure set, and then in the economiser (23), before being returned to the boiler drum.
12. The cooling water from the condenser is sprayed inside a cooling tower (1), creating a highly visible plume of water vapour, before being pumped back to the condenser (8) in cooling water cycle

13. The three turbine sets are coupled on the same shaft as the three-phase electrical generator (5) which generates an intermediate level voltage (typically 20 - 25 kV).
14. It is tested up by the unit transformer (4) to a voltage more suitable for transmission (typically 250 - 500 kV) and is sent out onto the three-phase transmission system (3).
15. Exhaust gas from the boiler is drawn by the induced draft fan (26) through an electrostatic precipitator (25) and is then vented through the chimney stack (27).

Thermal power plants require an uninterrupted supply of large volumes of water and are therefore located close to a river, lake or a reservoir. They require a small volume of highly pure demineralised and polished water for boiler make-up, auxiliary cooling and stator cooling and a large volume of relatively impure water for condenser cooling ash handling and other uses.

Filtered and chlorinated water is supplied to meet the requirements of the power plant and the colony.

Coal based power plants generate large amounts of ash (ash content in Indian coals is around 40%). Again large volumes of water are used for ash slurry pumping and for ash suppression. Boiler water treatment for high pressure boilers is almost always standardised.

Raw water is pretreated for removal of suspended solid, turbidity and organics in a pretreatment plant consisting of a clarifier and filter. Chlorine is dosed upstream of the clarifier for removal of organics.

A primary coagulant like alum is added in a flash mixer prior to the clarifier for coagulation of finely divide particles of clay, sand, algae. A polyelectrolyte is then added into the flocculating compartment of the clarifier for flocculation or agglomeration of flocs. Bulk of the suspended solids and turbidity are removed in the clarifier.

A granular media filter polishes the filtered water by removing the residual suspended solids and turbidity present in clarified water. Residual chlorine, if any is removed in an activated carbon filter. It is then demineralised and used as boiler feed make-up. An ultra filtration plant is also sometimes followed post demineralisation plant to remove colloidal silica, if any present in the demineralised (DM) water.

Corrosion due to dissolved oxygen is eliminated by thermal de-aeration followed by addition of hydrazine to remove the traces of oxygen and to keep a positive residual in the boiler feed water. Ammonia is also added to minimise condensate return line corrosion by carbon dioxide (CO_2). Morpholine, cyclohexylamine, etc. are generally avoided in high pressure operating boiler. Trisodium Phosphate is added into the boiler drum to maintain the pH at around 9.3 to 9.5. Continuous or intermittent blowdown of boiler water maintains the boiler salines within the limits prescribed by boiler manufactures. Effluents generated by power plants require to be treated either for safe disposal or for recovery of water for reuse.

Water can be recovered from effluents generated during regeneration of DM plants as well as cooling tower (CT) blowdown reused. Filter backwash water can be recycled back to the clarifier. Boiler blowdown can also be recovered and reused after treatment. A large quantity of water is used for ash handling, pumping the slurry and ash suppression. Ash retains only a small quantity of water and the balance appears as seepage. The effluents may contain SS & heavy metals and cannot be discharged without treatment.

Requirements of Water for each Application

Both the quality and quantity of water used in a power plant for the various applications in a power plant vary widely. Power plants require small quantity of high purity water and large volumes of relatively poor quality. Given below are the requirements of quality and quantity of water required in a power plant.

1. **Condenser Cooling:** Quantity of water required is around 130 m3/h per MW. Clarified water low in suspended solids and turbidity is used for cooling the condenser. As the quantity of water required for cooling is large, either once through cooling or open evaporative recirculating system involving a CT is used.

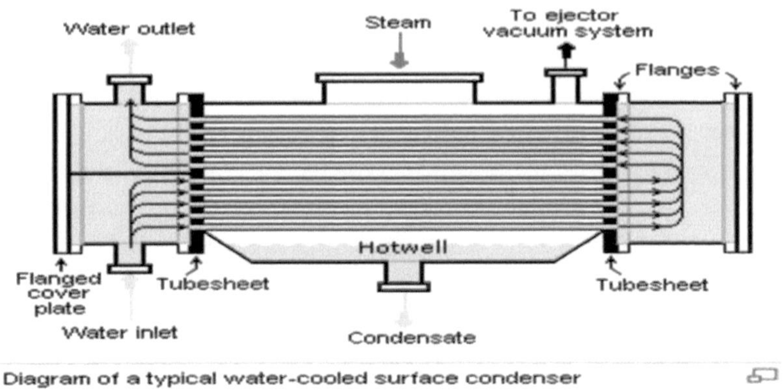

Figure 2

In case once through cooling is adopted, the only treatment given is continuous chlorination to contain biofouling. Corrosion is minimised by selecting superior metallurgy for the condenser.

Plants adopting the open evaporative recirculation cooling eliminate scaling either by using soft water as make-up or by acid addition to convert bicarbonates into carbonates. Plants in Maharashtra for example use soft water as make-up for the CT. Corrosion is minimized by selecting superior metallurgy for the condenser and limiting the cycles of concentration. Biofouling is contained by continuous chlorination.

2. **Auxiliary Cooling Water System**: Quantity of water is relatively small (around 10 m3/h per MW). The quality of water required is similar to that of condenser cooling.

3. **Stator Cooling Water System:** The system incorporates copper tubes and is subjected to high temperature and is prone to severe corrosion. Polished water is circulated through the closed circuit cooling system. Losses in the system is made up using approx. 2 m3/h polished water for a typical 210 MW plant.

4. **Boiler Water Make-up:** The system demands high purity water. Demineralised and polished water is used as make-up to compensate for the loss of water due to blowdown and other losses in the system. The quantity varies with the quantum of condensate recovered and it is unlikely to exceed 2% maximum of feed water flow for good systems.

5. **Hydro Ash Disposal:** Large quantity of water is required for pumping the ash slurry to the ash pond and for ash suppression. The quantity of water required will vary with the amount of ash (ratio as ash:water is normally 1:10 by volume).
6. **Drinking & Sanitation:** Potable water supplied must comply with the requirements laid down in IS 15000. Clarified and filtered water is chlorinated and distributed to all sections of the power plant. The norm is approx. 130 litre/day/person. It is possible to conserve water by installing electronically operated valves wherever possible. Online water purifiers are also installed at important locations like cafeteria, admin offices and laboratory.
7. **Miscellaneous applications:** These include water used for lubrication & sealing, chemical preparation and fire fighting system, make-up, etc. Clarified and filtered water is used. The quantity of water required is small and will depend on system losses and frequency of fire fighting drills conducted. Water used for fire fighting is prone to corrosion as water remains stagnant most of the time and it is customary to incorporate a corrosion inhibitor.

Contaminants Present in Water, Their Harmful effects and Methods of Removal

Sources of water:

Power plants require an uninterrupted supply of large volumes of water and are therefore located close to a river, canal, lake or a reservoir. Power plants located in coastal areas use sea water for most of the applications like cooling and polished desalinated water as boiler feed make-up.

Impurities present in water:

There are essentially two types of impurities present in water.

- Undissolved impurities
- Dissolved impurities.

Undissolved impurities of relevance to power plant operation include suspended solids (SS), turbidity, organic matter and colloidal silica. These are removed in the pretreatment plant. Dissolved impurities include hardness, sodium, alkalinity, chlorides, sulphates, nitrates, silica, dissolved oxygen and carbon dioxide. These are removed in the DM plant. Ultrafiltration system removes colloidal silica. Removal of oxygen takes place in the thermal deaerator. Hydrazine is added into deaerated water storage tank to remove the residual dissolved oxygen. Ammonia or an amine like Morpholine (pH above 8.5) is added to minimise corrosion of condensate return lines due to carbon dioxide.

Different processes of water purification are discussed now in detail in subsequent paragraphs, which starts from pretreatment plant followed by demineralisation, post treatment of DM water for boiler make-up, other uses, etc. as read under subsequent different chemistry chapters.

Water Treatment Plant

The water treatment plant generally consists of two major sections, viz., pretreatment plant & demineralisation plant

Pretreatment Plant

The objective of pretreating water is to remove impurities that are likely to affect the performance of the downstream DM plant.

The main stages of pretreatment are:

- Aeration
- Pre-chlorination
- Coagulation
- Flocculation
- Clarification
- Filtration and
- De chlorination

Aeration

Increase the dissolved oxygen concentration in water. Decrease CO_2 concentration thereby reduce its corrosiveness. Reduce taste and odour caused by dissolved gases such as hydrogen sulphide (H_2S) and methane (CH_4) that are removed during aeration. Oxidise iron and manganese from their soluble to insoluble states and cause them to precipitate so that they may be removed by sedimentation and filtration processes. Remove certain volatile organic compounds.

Aeration is typically carried out in a cascade aerator in concrete. Feed water enters the aerator from the central inlet pipe, flows upwards and cascades downwards in a free fall manner to a total depth of 2-3 metres with splashing to the extent of 5-6 times where reduction of iron and manganese content is required. Sometimes cascading is affected with multiple trays, usually containing coke, which has catalytic effect on precipitating these. Pre-chlorination of the raw water is also often used to accelerate the rate of oxidation.

Cascade Aerator

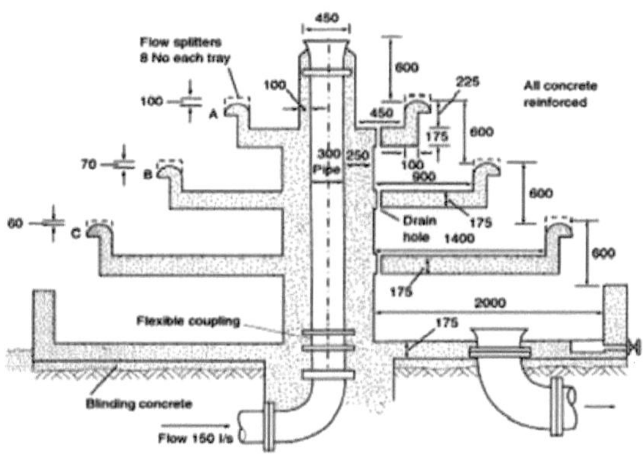

Figure 3

For increasing the efficiency of aeration, cascade aerator is used. In this, water flows out of the top outlet and cascades down while coming in intimate contact with air. The design of cascade aerator ensures appropriate contact time of water and air. This is usually followed for stripping of too much carbon dioxide (as in DM plant) and/or volatile organics like chloroform as well as H_2S, NH_3, etc. are required. Equally efficient to this type of aerator is spray type nozzle type aerator, (generally spraying in a 'Pond', the nozzles commonly of 2.5 to 4 cms diametre and discharging about 5-10 l/sec in the form of jets and hitting metal plates to produce a fine spray exposing countless droplets of water to the atmosphere. Thus the efficiency of CO_2 removal from water fall aerators has usually been noted as below:

Sprays	90%
Trays	70%
Cascading	50%

Yet, adopting the cascading type of aerator is found as first choice, since reduction of CO_2 up to 10 ppm is normally considered satisfactory. Another diffuser called 'Air diffusion aerator' is a primary biological system reducing biomass from water by air oxidation/activated sludge treatment of municipal and industrial waste, providing maximum water surface per unit volume of air. The velocity of bubbles ascending through water is much lower than the velocity of free falling drops of water, providing a longer contact time. The aeration tank is mostly vertically divided into two compartments by a baffle extending from the point of air introduction to a point above the tank floor. Large volume of low pressure air admitted through aeration grid creates an air lift, causing mass circulation around the baffle. This promotes maximum gas/liquid contact & optimum mixing and desired BOD removal is above 90% promoting surface aerators as most common in use for reclaiming city sewage as industrial water including for some TPS.

Pre-Chlorination

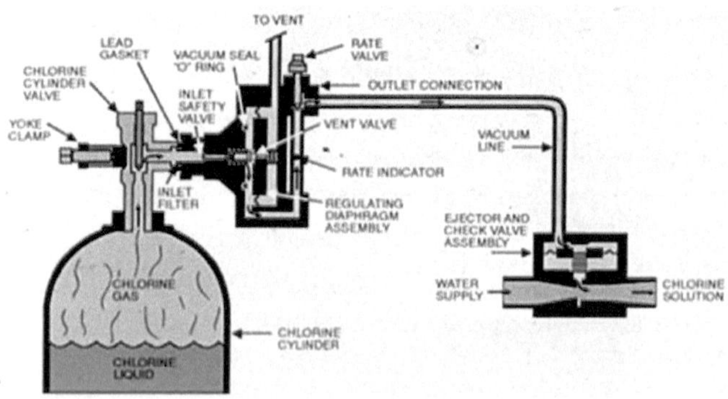

Figure 4

The objective in chlorinating the water is primarily to oxidise organic matter and prevent microbial growths. Up to 2-5 mg/l of continuous chlorination is the normal practice. Establishing chlorine demand and adjusting the chlorine dose is the best option. Chlorination is carried out using gas chlorinator which dissolves chlorine in water. When chlorine gas is dissolved in water, a mixture of acids - hypochlorous and hydrochloric is formed. Hypochlorous acid dissociates to form hypochlorite ion which is a powerful oxidant. Chlorine is effective in a narrow range of pH (6.5 to 7.0). Pre-chlorination also oxidises iron/manganese salts and helps them in easy settling.

Coagulation

While most particles settle on their own, finely divided particles of clay, sand or algae do not readily settle. These particles carry negative surface charges which prevent them from colliding with each other. Addition of a coagulant like alum neutralises the surface charge allowing the particles to coalesce with each other and become a larger particle called flocs. Thus, freshly precipitated aluminium and/or iron hydroxides which have electro-positive charge are added to it. The charges on the colloids causing turbidity are neutralised and the process is called coagulation.

Alum reacts with the alkalinity of water and forms insoluble $Al(OH)_3$ precipitate and H_2SO_4 acid immediately reacts with the alkalinity of water and forms salt and water.The aluminium cation neutralises the negative charge on the colloidal particles. In addition, the precipitate is fluffy in nature and helps in adsorbing the fine particles. The process is pH sensitive and works best in the narrow range of 6 - 7.5. Addition of alum results in a decrease in alkalinity and increase in sulphate.

$$Al_2(SO_4)_3 + 6\,HCO_3^- \rightarrow 2\,Al(OH)_3 + 3\,SO_4^{2-} + 6\,CO_2$$
$$FeSO_4 + 6\,HCO_3^- \rightarrow 2\,Fe(OH)_2 + 3\,SO_4^{2-} + 6\,CO_2$$

The equipment used for coagulation is a flash mixer fitted with a high speed mixer. It helps in mixing the coagulant with the incoming water. The retention time in the flash mixer is typically one to two minutes.

Sodium aluminate also alone reacts with water to give aluminium hydroxides.

$$Na_2Al_2O_4 + H_2O \rightarrow 2\,Al(OH)_3 + 2\,NaOH$$

A number of coagulants are in common use in water treatment, including ferric sulphate/chloride. They are generally used for coagulating water having high organic matter. The requirements for effective coagulation include optimal pH, sufficient alkalinity choice of coagulant, retention time and high speed agitation. Optimal coagulation & minimum floc solubility with use of alum occurs at pH 6.0 to 7.0 for optimum turbidity removal and pH 4.5 to 5.5 for colour removal. Minimum residual aluminium is found at pH 6.3 and minimum iron residual at pH 5.4 and then again at pH 9.3 to 13.

Flocculation

The term flocculation refers to the process of agglomeration of flocs and this is aided by the addition of a polyelectrolyte (or a coagulant aid). Polyelectrolytes are long chain compounds and the flocs get attached to them.

The equipment used is a flocculator fitted with slow moving paddles providing a gentle motion that promotes the growth of the flocs in size and weight. Polyelectrolytes are cationic, anionic or non-ionic. The choice will depend on the application. In normal water treatment applications, cationic polyelectrolytes are widely used. The dose of coagulant/polyelectrolyte for suitable clarity of water is established by laboratory jar test.

Jar Test Apparatus

Figure 5

The following factors generally influence coagulation:

1. Water Quality
2. Quantity & characteristics of colloidal matters
3. pH of water
4. Time of fast mixing & flocculation and speed of paddles
5. Temperature and alkalinity
6. Characteristics of ions in water

The turbidity measurement alone is inadequate to measure effectiveness of coagulation. It is the particle size, which matters most. Based on the turbidity & alkalinity of water, amendability to coagulate can be judged as below:

Raw Water	Quality	Amenability
Type I	High Turbidity (>100 NTU) High Alkalanity (>250 ppm)	Easiest to coagulate
Type II	High Turbidity Low Alkalinity (<50 ppm)	Alkalinity addition needed, if pH drops
Type III	Low Turbidity (<50 NTU) High Alkalinity	Large dose of alum promote precipitation of Al (OH). Also, coagulant aid & dose needed
Type IV	Low Turbidity Low Alkalinity	Most difficult to coagulate alkalinity addition is needed to produce Type III water or clay addition to produce Type II water

Low TDS water and small particles favour high particle surface charge (-ve in nature). This also depends on composition of mineral salts present in water. Hence increasing TDS is one of the means to reduce surface charge for settling purposes of floc. Optimal pH range for flocculation with aluminium sulphate is 5.5 to 7.5, with ferric sulphate 8 to 11, and with ferric sulphate for colour removal is 5.0 to 6.0. But for highly coloured and soft water pH is 3.8. A colour of 50 Hazen units comprising of Humic & Fulvic acids may require 20 to 25 mg/l of alum, but when iron is present as non-reactive organic complex, it is necessary to chlorinate first before coagulation with alum & sodium aluminate, the optimal pH range, being at pH 6 to 7, at which maximum organic removal, minimum aluminium residual & max. floc density are obtained. Actually when pH drops below 6.2, residual aluminium in solution tends to increase. Though experts define coagulation & flocculation differently for different purposes, in the field of water treatment plant, it is customary to reserve the term coagulation for destabilisation of electrical charges on colloids, and the term flocculation for agglomeration of discharged colloids.

For the later process, it is not only the aluminium ions in solution which react with the clay particles, but it is the mass of rapidly precipitating & flocculating $Al(OH)_3$ which enmeshes them. The precipitating $Al(OH)_3$ has a weak (+) charge, which will enhance the initial deposition of the hydroxides on to the surface of the clay particles. Hence, for these agglomeration purposes, use has been made of natural polymers alone or combined with mineral salts (Nirmali seed powder, potato starch, alginates, dextrins, gelatine, activated silica, etc.) and later synthetic polymers of high molecular weight, soluble in water as coagulating agents.

Coagulation aids – later as polyeloctrolytes, improve floc properties formed by coagulant & enhance coagulation. Among the coagulant aids are clays, activated silica, bentonite and different polymeric compound, which prevent light weight, fragile, slow settling floc consisting primarily of $Al(OH)_3$ or caused by water low in turbidity.

Though amount of coagulant dose is fixed experimentally in the laboratory, but theoretically corresponds to a zeta potential of about -3mv to -4mv or practically close to zero. For colloids in natural water in a pH range of 5-8, the zeta potential is generally - 14 to 30 mv. Zeta potential in short can be defined as potential difference between the wall of the particle and the bulk of the liquid. Its value determines the extent of the electostatic force of repulsion between the colloidal particles & hence their probability of adhesion. The coagulant generally reduces the zeta potential.

The zeta potential is not completely cancelled by using non-ionic polymers (polyacrylamide of molar mass 1 to 30 million) & anionic polymers (polyacrylamide partially hydrolysed by soda—contains both groups permitting adsorbtion & negatively ionised groups—carboxyl or sulphuric groups which extend the polymer and molar mass of several million). They modify the surface state of the particles & their zeta potential causing them to agglomerate into flocs.

Cationic polymers however having (+)ve electrical charge due to presence of amine (Polyvinylamines), imines (Polyetheleneimines) or quaternary ammonium groups & having molar mass of not exceeding one million cancel the zeta potential to zero at a lower dose of 4 ppm maximum, and is classified in order of efficiency.

Cationic polymers are the most suitable type for town sewage sludge or where sludge is rich in organic matter. While anionic & non-ionic types are suitable for sludge containing colloidal metal hydroxides or sludge rich in clay, kaoline or activated carbon powder. A combination of lime and anionic polymer flocculation can work very well on highly colloidal organic sludge. Anionic polyelectrolyte is also used to increase the floc bulk of metallic hydroxides when used as oil adsorbent helping oil removal easily in sludge.

Anionic & non-ionic polymers are generally ineffective as primary coagulant due to (-)ve charge on it, since silica (SiO_2) the main ingredient for causing turbidity is also having (-)ve charge. However, these are effective as coagulant aid, by bridging mechanism after destabilising of colloids with the help of extraneous electrolyte such as NaCl, $CaCl_2$, alum, etc. particularly bivalent & higher salt.

Polymer will not work alone in case of

1. Low turbidity & high alkalinity
2. Low alkilinity & low turbidity.

Coagulant aids such as clay should be added ahead of polymer. Polymers are not acidic & do not lower the pH of water as alum does. Dilute polyelectrolyte solutions lose activity with time & at high temperature and hence solution should not be prepared for more than 24 hours consumption.

Flocculation & Sedimentation

The purpose of flocculation is thus to cause the particles either to agglomerate into large floc particles which will settle rapidly under gravity and can therefore be separated from the water by sedimentation, or to grow in size sufficiently to ensure that they will be captured or retained suitably on or within the bed of a filter unit. When treating water with a considerable concentration of suspended solids or organic matter & requiring correspondingly higher dosage rates of coagulant, it is often necessary to incorporate a sedimentation stage ahead of filtration in order to reduce the load on filter units. This can be understood like this: By considering raw water with a turbidity of 100 NTU (approx. 100 mg/l on silica scale). The raw water comprising this turbidity has a specific gravity of about 2.5 and therefore occupies about 40 ppm by volume.

The addition of a coagulant such as aluminium sulphate in a dosage of perhaps 20 mg/l results in the entrapment of the suspended particles in the gelatinous floc produced. It has not been generally recognised that the volume of material produced by flocculation is in the order of 5000 ppm. Such a volume of material cannot be applied to a rapid sand filter of the traditional type without rapid clogging. Therefore, a period of sedimentation is required to reduce the volume to 1000 ppm or less in order to permit the filters to function for an adequate period of time between cleanings. This sedimentation stage is accommodated with clariflocculator – commonly called as clarifier. In this case, the flocculation conditions must be arranged so that large preferably high density floc particles are produced, which depend on various factors including initial concentration of particles present as floc colloids, the condition of agitation (magnitude of the shearing action) retention time in the reaction vessel & ageing to form sludge, etc.

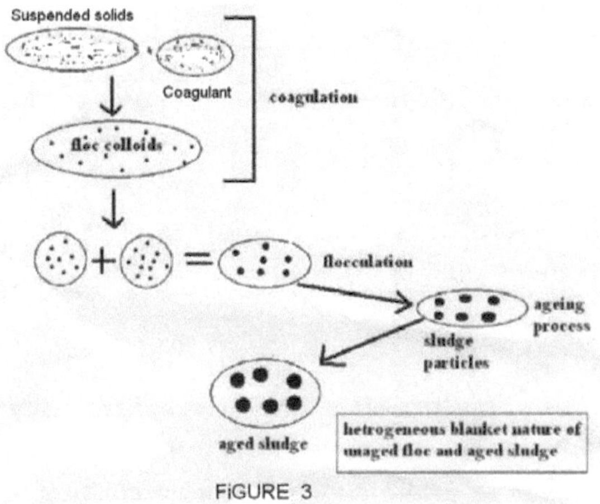

Figure 6

The coagulant chemicals normally employed are salts of Al & Fe (ic), commonly referred to give their hydroxides to have surface charge for destabilisation of colloids, but actually a very large range of polymeric hydrolysis products also have been found to exist depending upon the treatment conditions applied. These products have often been recognised as [$Al(OH)_3M$] configuration, in which M is another anion such as SO_4 or Cl absorbed from the environmental water in which they are formed. They are hydrophilic & contain a considerable amount of absorbed water; the degree of which depends on degree of crosslinking. If the floc is highly cross linked, it will retain less structured water, have high mechanical strength & high density - will be able to withstand shearing conditions to a greater extent and be capable of achieving greater size in combination with floc colloids to finally give a high settling rate. Maximum crosslinking will occur when coagulant chemical dose is low relative to the amount of solids available for coagulation. This therefore indicates that coagulant dose should be optimum. Any overdose of this will affect crosslinking to produce weakly crosslinked, swollen low density flocs with higher water regain to collapse soon.

Further, high shear mixing is especially important when cationic polymers are used as primary coagulant and hence it is advisable to feed them as far ahead of clarifier. However, when coagulant aid is used, there is need to avoid high shear mixing to prevent interference with the polymer's bridging function. Only moderate turbulence is needed to generate floc growth. Hence, coagulant aid must be added sufficiently to the downstream point of application of primary coagulant preferably near the sludge blanket zone. Since flocculation with polyelectrolyte is immediate, the injection point most often is located almost immediately ahead of dewatering unit, the mixing being violent but short. If both cation & anion forms are to be added, the two injection points should be set separately on the feed pipe at a distance correspondingly to a flow time of 1 to 2 minutes.

The time that must elapse between the addition of coagulant (Al salt) & flocculant (polyelectrolyte), activated silica, etc., is of great importance & must be determined by lab test. Similarly, suitability of flocculation aid is also determined by lab test. Reaction time for flocculation is usally 20 - 300 seconds by injecting a solution of concentration of 0.5 to 1.0%.

Any pH conditioning chemical acid or alkali must be separately rapidly mixed and dispersed in the water, with a view to achieve optimum pH for final coagulating effect. Lime should be added after the coagulant to avoid interaction of lime with alkaline hardness of water & consequent formation of colloidal $CaCO_3$, which has a dispersant effect on the high pH iron hydrolysis product (pH 8).

Conventional Clarification Equipment

This mainly comprises of coagulation, flocculation and sedimentation processes and have three distinct features:

1. High shear, rapid mix for coagulation.
2. Low shear, high retention time, moderate mixing for flocculation.
3. Liquid & solids separation for sedimentation.

Originally, conventional clarification units consisted of large rectangular concrete basins, divided into 2 or 3 sections, each stage of clarification processes occurred in a single section of the basin. Water movement was horizontal with plug flow through these systems. But in order to achieve high throughput efficiency, compact and relatively economical upflow clarifiers have been the most conventional units till date to maintain high clarity effluents. These are characterised by increased solids contact through internal sludge recirculation basins much smaller in size than horizontal flow basins. These also maximise the linear overflow weir length, while minimise the opportunity for short circuiting through the settling zone. The sedimentation tank is generally designed for a depth of 4 or 2 metre corresponding to overflow rate of 2 m/h & 1 m/h respectively.

The surface area of the tank is determined on the larger of the areas on the basis of the following:

a. Surface loading in m^3/m^2 per day
b. Solid loading in kg of suspended solids/m^2 per day

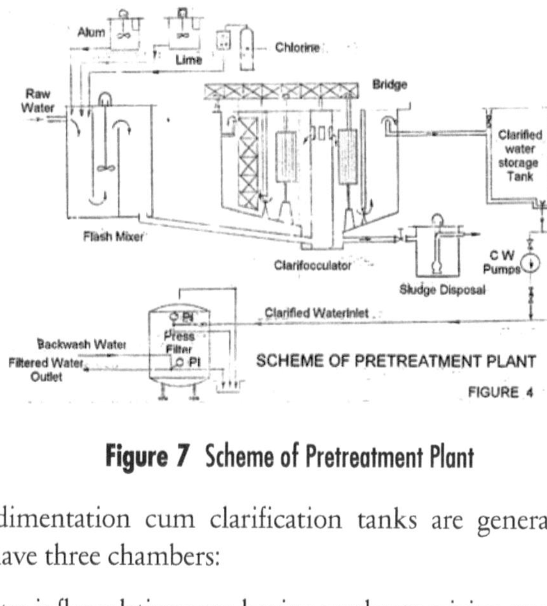

Figure 7 Scheme of Pretreatment Plant

The sedimentation cum clarification tanks are generallly circular in nature and have three chambers:

a. The centre is flocculating zone having moderate mixing arrangement.
b. The middle (the larger size) is both settling and flocculating zone having very slow rotating paddles & sludge scrapper blades fixed with a rotating bridge.
c. The outermost chamber is for clean water overflow. The middle chamber also has a sludge pit with sludge disposal arrangement. Over & above, before this sedimentation/clarification tank water has to pass through a flash mixing arrangement in which the coagulant chemicals are mixed violently and with very less retention in it, this mixed water rapidly is passed on to the central chamber of the clarification tank. The typical data of a conventional clariflocculation tank are generally as below:

Flash mixer retention	1 minute
Flocculation time	30 minutes
Clariflocculor tank retention	3 hours
Tank side water depth	3 to 4 metres
Floor slope (at conical portion)	1:12
Flocculation paddle speed	40 - 50 rpm
Scrapper/bridge revolution	one revolution per hour (approx.)
Weir loading (flow per unit length)	not greater than 25 m^3/h/m

(Throughput efficiency is achieved by maximising the linear overflow weir length while minimising the opportunity for short circuiting through the settling zone. In addition, the two mixing stages for coagulation & flocculation occur within the same clarification chamber). Experiences & research have shown paddle speed, followed by flocculation time & alum dosage to be the most significant independent variables.

It seems to be a good rule to provide a storage sludge depth of about 30 cm near the outlet and a sludge storage depth of 2 metre or more near the beginning of the zone of settling. In water, sludge will not usually move down flatter slopes under the effect of gravity. Hence, hopper bottoms are generally used with slopes approximately 55o for sludge blanket clarifiers and 5o for solid contact clarifiers.

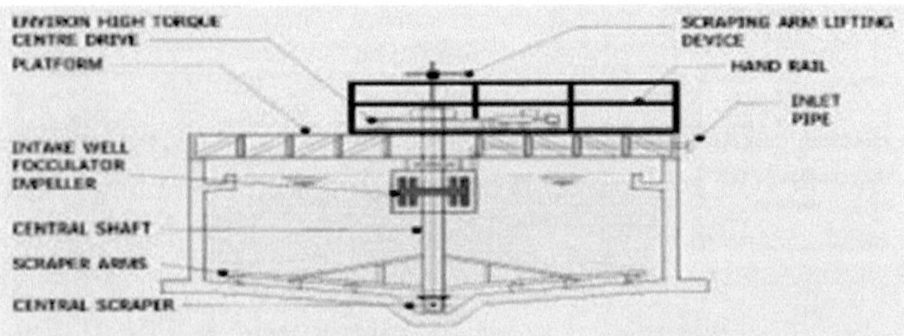

GENERAL ARRANGEMENT of CLARIFLOCCULATOR

General arrangement of a typical clariflocculator is shown as below:

Apart from the above, conventional type of clariflocculator, other types are also in practice, few of which are briefed as follows:

Sludge Blanket Clarifier

Most current high rate sedimentation clarifiers are based upon the sludge blanket or solids contact concept. In sludge blanket operation, the 'floc colloids' produced are immediately introduced into the base of the separating zone to flow upwards through a fluidised bed or sludge blanket (formed by a high volumetric concentration of aged floc particles & accommodated from earlier treatment). It is important that the material entering the sludge blanket zone consists solely of floc colloids & virtually no large, low density floc particles – which in turn will not adhere to the aged sludge particles (in view of saturation of its size enlargement), but will pass through sludge blanket & be carried over by the clarified water. The volume of sludge blanket zone needs to remain constant and any surplus has to be removed corresponding to the input volume of floc colloids. Also, there is need for appropriate coagulant dosing; excessive amount will have adverse effect on ageing time in the sludge blanket. The sketch of above clarifier is illustrated as below:

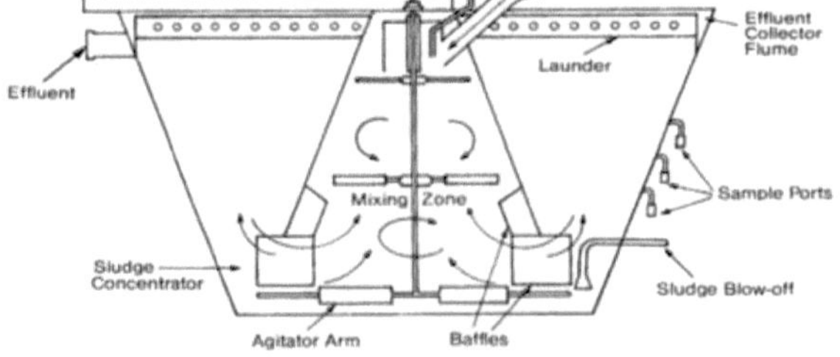

Solids Contact Clarifier – Reactivator

This concept comes actually from the lime softening process in which precipitation of $CaCO_3$ from supersaturated solution of calcium carbonate formed by the reaction of lime & bicarbonate hardness was found to be greatly accelerated by contact with previously precipitated $CaCO_3$. Also, silica reduction requires the use of a solid contact type clarifier because of the high sludge volume and detention time required for good removal.

In this clarifier (as shown in figure), special arrangement is made in the central sludge collection pit (cf: conventional clarifier). By means of a draught tube positioned centrally above the sludge pit & already equipped internally with a low head pump impellor or an hydraulic ejector, part of the sludge from pit is drawn up to mix it with the incoming water & treatment chemicals. On leaving the top of draught tube, the water & sludge are redirected downwards, pass under the central skirt containing the sludge recirculation zone & enter the separation zone.

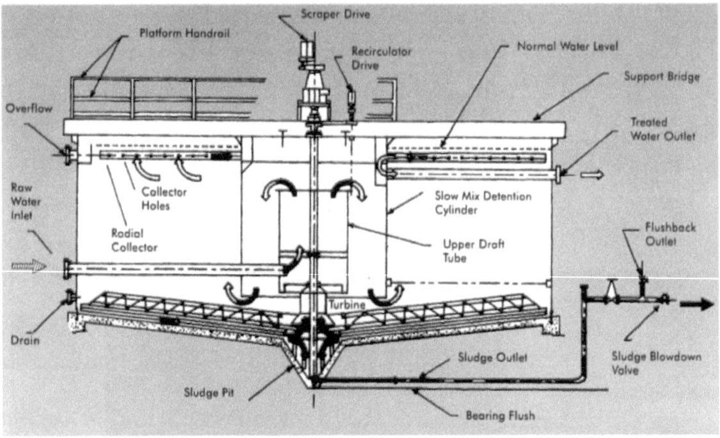

Solids contact clarifier

Lamella Clarifier

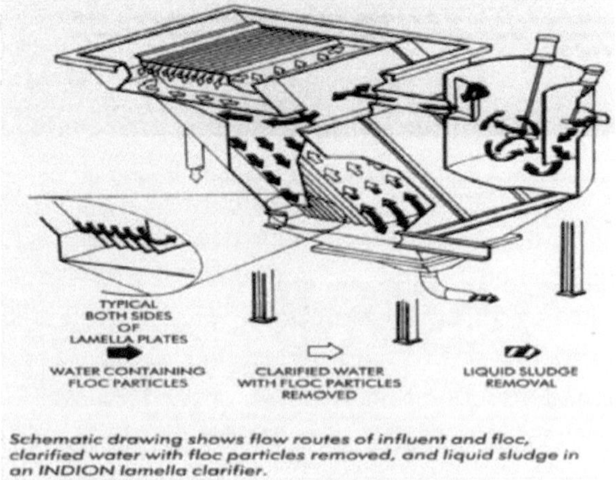

Schematic drawing shows flow routes of influent and floc, clarified water with floc particles removed, and liquid sludge in an INDION lamella clarifier.

Lamella clarifiers are suggested for high turbidity (starting from 300 NTU and above). For low turbidity water, a separate recirculation system should be provided. An advantage of this clarifier is the increased amount of settling area with small foot print.

A lamella type clarifier consists of a large number of inclined surfaces in the form of plates made usually of FRP. They are arranged in close proximity & inclined at an angle of 55°, which divide the lamella tank into several integral section/ chambers. Only one side of each chamber (i.e. top surface of a plate) provides a settling surface. The 'pretreated' feed stream enters the Lamella tank & transverses through feed ducts longitudinally along each side of the Lamella plates through a bottomless distribution duct. The liquid/solids feed stream then enters each plate chamber near the bottom section of the plates & flows upward between them with solids continuing to slide down the plate surface finally to a collection hopper. Providing the optimum length to width ratio of plates will help maintain ideal condition for separation. Arrangement of plates are made in such a fashion that a single operator can take them out for inspection & clean sticky flocks on it, if any, then reinstall them.

Tube Settler

Tube settler emerges as another shallow depth sedimentation clarifier like Lamella plate type - the difference here is tubes are in place of plates and the tubes are generally inclined at a minimum of 60° to ensure self draining. The design loading rates may however be a bit low versus that of a Lamella, in view of ratio of settling area to plan area being relatively lower. More projected settling area surfaces are not available here for settling due to random distribution and collection arrangement in these types of shallow depth sedimentation.

Operation Of Clariflocculator

Before operation of a clariflocculator, it has to be ensured that:

a. Raw water quality along with lab jar test records is known.

b. All coagulant chemicals including aid, if any, are available in form of solution of required strength. For alum solution having concentration of 10%, anionic, cationic and non-ionic polyelectrolyte solution of 0.5 to 1% concentration is used. Saturated lime solution has also to be kept ready, but to be dosed only when pH adjustment found necessary.

c. Chlorine dosing equipment without any leak has to be kept ready and to be dosed as assessed under lab jar test.

d. Equipments, valves, flow metres, etc. are found in order.

e. Feed water (raw water) should be dosed with adequate amount of chemicals as per the jar test results.

f. Sludge has to be removed without much accumulation in the clarifier.

g. After start of the operation, the clarified water needs to be tested for turbidity, pH, residual chlorine, organic matter, colloidal silica, iron, manganese, etc. and see whether the quality is available as per design parametres atleast after 2 - 3 hours of operation.

h. Generally clarifiers in a thermal power plants are designed to furnish turbidity of less than 10 NTU, pH in the range of 6.5 to 7.0, residual chlorine 0.2 ppm & iron and manganese along with other heavy metals not more than 0.01 ppm.

i. The following are the impacts of chemicals dosed on the original quality of raw water. Each of the chemicals of 1 ppm dose changes the quality of raw water as below:

j. Polyelectrolytes have been used both as a coagulant and as a coagulant aid to increase the efficiency of the removal of turbidity & colour and also to remove some of the new pollutants not affected by conventional coagulants.

Anionic & non-ionic polymers are generally ineffective as primary coagulants due to (-)ve charge on it and SiO_2 is also having (-)ve charge. But these are effective as coagulant aids by bridging mechanism after destabilizing of colloids. Anionic polymer's effectiveness can be increased when used in conjunction with an extraneous electrolyte such as NaCl or $CaCl_2$ or a coagulant such as alum. Overdosing of anionic polymer can seriously inhibit coagulation and must be avoided. Use of only 0.1 to 0.5 mg/l of anionic polymer in conjunction with alum can improve floc settleability & toughness significantly.

Dosage of only 0.5 to 1.5 mg/l of cationic polymer is often effective for coagulation, whereas as much as 10 to 20 times that amount of alum is required for same results.

Polymer will not work alone in case of (1) low turbidity & high alkalinity (2) low turbidity & low alkalinity. Coagulant aids such as clay should be added ahead of polymer.

If both cationic & anionic polymers are used simultaneously, the two injection points should be set separately as the feed pipe at a distance corresponding to a flow time of one to two minutes.

k. Solid contact clarifiers do not function well as liquid-solid separators when treating raw water of low turbidities.

l. Clariflocculator having good mixing & flocculating arrangement and if free of short circuiting can produce results approx. three times better than achieved by lab jar test. Apparently, this is due to continued flocculation taking place under gravitational influence, a phenomenon called 'sweep flocculation'.

When raw water turbidity is rising, the alum dosage for a given turbidity is frequently smaller than at the same turbidity level when turbidity is diminishing, apparently because of differences in suspended particle size. Similar day to day deviations in settled water turbidities occur due to such factors, including density current effects. The ratio of 12 to 15 between turbity vs coagulant (alum) dose for usual surface water has been experienced with still lower dose of coagulant having raw water turbidities of more than 500 NTU.

m. Finally clariflocculator operating results can be affected by many factors:

   a. Amenability of the raw water to coagulate & flocculate.

   b. Design of the plant facilities for coagulation, dispersion, flocculation and sedimentation including the short circuiting characteristics of the basins provided.

   c. Surface loading of settling basins.

   d. Tightness of the control of the coagulation process by the operating personnel.

Effect on Water Composition (With addition of each ppm of pure chemical)

	ppm $CaCO_3$ change						
Chemical	Molecular weight	M Alk	CO_2	SO_4	Cl	Ca	Na
Alum	666	- 0.45	+ 0.45	+ 0.45	-	-	-
Sodium Aluminate	164	+ 0.61	- 0.61	-	-	-	+ 0.61
Ferric Chloride	270	- 0.55	+ 0.55	-	+ 0.55	-	-
Ferric Sulphate	562	- 0.53	+ 0.53	+ 0.53	-	-	-
Ferrous Sulphate	278	- 0.36	+ 0.36	+ 0.36	-	-	-
Lime	74	+ 1.35	- 1.35	-	-	+ 1.35	-

e. Water quality goals of those responsible for the plant operation. Nevertheless, there is no ideal clariflocculator completely free from practical difficulties in obtaining the desired results continuously, and some times the operation remains more an art than science.

Filtration

In view of the natural water having mostly colloidal matters in the form of turbidity, which need to be removed, under thermal power station practices, the filtration is the second stage in purification of water in which remnant of impurities of suspended matters from clarification plant is removed to the extent of 98 - 99%. Mainly two types of filters are in convention, viz., gravity sand filters & pressure sand filters.

Dual media/multimedia filters are further in practice as per requirement of removal of suspended solid and outlet quality of water. Other media alone like anthracite in place of sand is also used to remove residual precipitated hardness salts remaining after clarification under softening precipitation. Apart from river sand & anthracite, Quartz sand, silica sand; garnet, magnetite & other materials are also used as filtration media particularly in multi bed system. A more recent development using multimedia beds is contact flocculation, in which coagulant is applied directly at the inlet of the rapid filter unit, in which quick mixing followed directly by filtration technique is employed. This is however applicable for a very low turbid water e.g. ground water, poorly filtered town water supply, etc., and is based on the principle that a brief high intensity mixing can result in the production of floc volumes in raw water in the order of less than 1000 ppm, an optimum value generally recognised not to clog the rapid filters.

Rapid Gravity Sand Filters

Essentially these consist of (a) Feed water having depth of 1 to 1.5 metre which provides a constant head of water to overcome the resistance of the filter bed & thereby promoting downward flow of water through the sand bed. (b) A bed of graded sand of about 1.2 metres & having effective diametre of 0.15 to 0.35 mm — the sand should be clean, free from clay & organic matter and preferably rounded. This sand bed is further supported by a layer of gravel & coarse sand 0.30 to 0.40 metres preventing the fine grains being carried into drainage pipes. (c) Underdrainage system at the bottom of filter bed consisting of porous or perforated pipes serving as outlet of filter water, support of filter bed and allowing backwash with entry of filtered water/air. (d) A system of flow control valves. A typical type of slow sand filter is illustrated in Fig.9

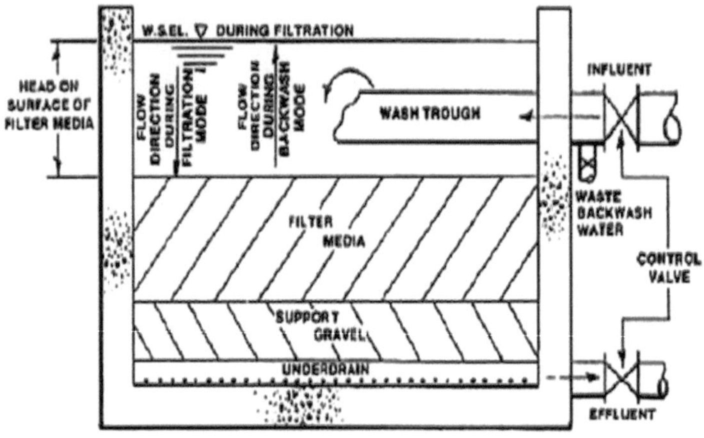

GRAVITY FILTER MODULE

Depending on the clogging of sand bed with organic matter, which gives resistance to flow & when backwash is not helpful after few months of operation, the sand bed (after draining the supernatant water is cleaned by scrapping off the top portion of sand layer to a depth of 1 or 2 cm by skilled labourer. But even after a few operation, if the thickness of sand bed is reduced to 0.5 to 0.8 metres, then the plant is needed to have totally new bed. The construction of rapid sand filters of gravity type are almost same except the following:

1. Rate of filtration is here 5 - 8 m/h in place of 0.1 to 0.4 m/h in slow sand, and thus, the rapid type occupies comparatively less space. Reactive siize of sand in the rapid type is 0.45 to 0.55 mm in place of 0.15 to 0.35 mm as in slow sand filters. Rapid gravity filters with single media or dual media (sand & anthracite) are used for large flow rates.

Typical rapid sand filter design of conventional type

Filtration velocity	4 to 8 m/h
Sand quality	Generally, river sand free from clay, dust, etc.
Effective size of sand	0.45 to 0.55 mm
Sand bed depth	600 - 700 mm
Silex and gravels (supporting media)	400 mm
Level of water above the top of bed	0.9 to 1.6 m
Total area of openings in the underdrain laterals of perforated pipe system (for even distribution of backwash water through out the filter bed)	Computed based on flow rate
Velocity of backwash	25 to 37 m/h
Compressed air if used during backwash	$0.66\ m^3/m^2/h$ @ $0.4\ kg/cm^2$
Free board	50 to 75% of filtering bed depth

Pressure Filters

With the same principle of underlying the filter beds of rapid type, these are enclosed in pressure vessels to facilitate higher flow rate, but by reducing the filter sand bed depth to 0.5 to 0.8 metre and keeping filtering rate 8 to 12 m/h, which can remove turbidity, suspended solids very efficiently. The greatest utility of these filters are that they occupy little space in comparison to slow sand filters & with dual or multimedia filters these can be operated up to 15 m/h filtration rate without compromising the quality. Filter sand used to remove oxidised iron & manganese are usually coarser (effective size 0.6 mm) than those used with coagulated water. In dual media filters, the use of 0.95 mm anthracite was found to produce more favourable operating conditions than anthracite of 0.70 in conjunction with the use of 0.45 mm effective size sand.

Pressure filters frequently encounter mud ball problems or cementing of sand grains by deposition of iron, manganese & calcium. A well designed backwash system can aid in preventing this problem. Pressure filters are designed vertically or horizontally & have cylindrical steel shells & dished heads. Horizontal filters are also separated into compartments to allow individual backwashing. Vertical filters usually have up to 3 metre diametre and the horizontals have up to 2.5 metre diametre & 7.5 metre length, the capacity (flow) having nearly double of the former.

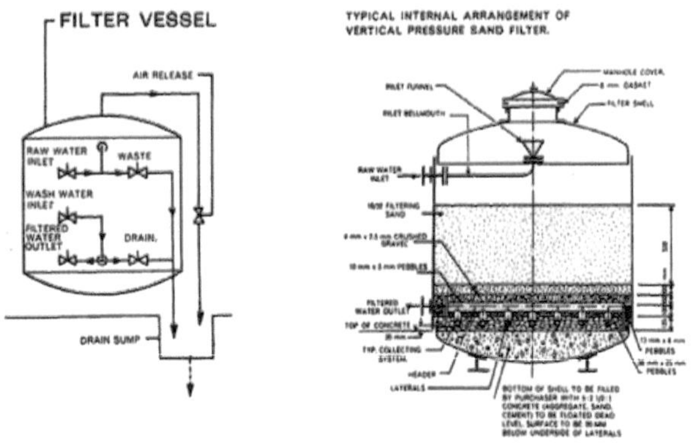

Vertical Pressure Sand Filter

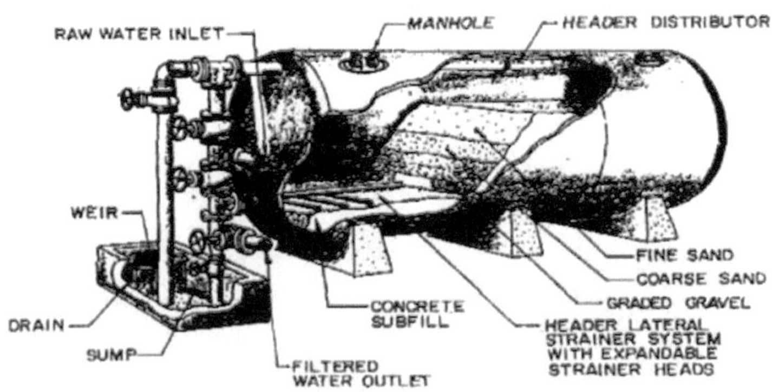

Horizontal Pressure Sand Filter

Other Filters

When treating water with a fairly low concentration of suspended solids and requiring only small amounts of coagulant chemicals for particle destabilisation, the following types are usually in practice.

a. **Precoat Filters:** These are particularly beneficial for oil particles & bacteria removal. Here the filter media typically diatomaceous earth forms a cake on a permeable base or septum. Filter cloth, porous stone tubes, porous paper, wire screens & wire wound tubes are also used as base materials. The supporting base material is first precoated with slurry of precoat media. Body feed then creates additional coat. These types of filters are however suitable for relatively small quantities of water and for specific purposes. In condensate polishing, the precoat is made of powdered ion exchange resins, which not only filters the corrosion products, but purifies the condensate in respect of unwanted ions. The precoat media is a consumable item. Filtration in presence of precoat media gives more pressure loss compared to regular media filters.

   Also, once a filter run has started, flow must continue or there is danger that some precoat may fall off the support. If the flow is intermittent, an automatic recycle pump may have to be installed.

b. **Upflow Filters:** For relatively low flow capacity and low filtration rate of 1.5 to2.5 m/h, upflow filters are also in use, which contain a single filter medium usually graded sand, fine at the top and gravels embedded as fixed at bottoms. This helps removal of coarser impurities of water at bottom, while smaller solid particles are retained further upward in the media.

c. **Varivoid Filter:** These are in line of contact filtration and are applicable to filter water having low concentration of impurities, but with usual high flow rate capacity. It is rather possible to use these filters as flocculators also & eliminate separate sedimentation equipment. In this filter, the porosity of sand bed is adjusted by using a binary mixture of sands, such that a small amount of fine sand is entirely retained within some of the pore spaces of the bulk coarser sand. This arrangement of the bed increases the tortuosity, enhance velocity gradients within the pore streams with localised shearing condition, which flocculate the small particles of suspended matter (if coagulant chemical mixed water traverses) as they travel, though the bed with growing size of floc to be retained further by bed.

d. **Automatic Valveless Filters:** These filters (figure 10) operating on the loss of head principle have been widely used as side stream filtration of cooling tower water by which overall turbidity in the cooling water is greately reduced. On these filters, there is nothing to watch to start & stop, and hence no manpower to control resulting with zero operating cost.

These filters comprise of a vertical steel tank divided into three sections:

1. Backwash storage space to hold the adequate amount of backwash water. It always fills to the same height providing uniform backwash rate and volume. Thus, the backwash stops when the level drops below the end of siphon beaker with the level dropping backwash rate starting from 44 m/h slows down to 30 m/h.
2. Filter bed compartment contains fine filter sand, disc type plastic strainers uniformly collect the filtered water and distribute the backwash water without any gravel layer.
3. Collector chamber for the effluent, which is directly connected with the backwash storage space. The filterate flows through a single vertical duct which is vented at the top.

In view of these filters generally operating on head loss of 0.12 to 0.15 kg/cm^2, it can just be called as 'water washer' and the filterate at times may contain up to 5 ppm of suspended solids. Benefits of these filters are many. Still these have not been widely used for pretreating water for DM plant operations. These have also been extensively used in removing undesirable solids and turbidities out of waste water.

Operation of Filters

While operating a conventional filter, the following have to be kept in mind:

a. Before operation, air has to be released both from air release valve & filter outlet valve.

b. Backwash has to be conducted at specified flow rate (usually at 30 - 35 m/h till turbidity of inlet & outlet water is found same. After backwash, clarified water has to be put on the filter bed in a fashion to avoid turbulence, with a slow start and bringing to full rate over a period of 15 - 30 minutes.

c. Backwashing of the filters are done when pressure drop across the filters exceeds 0.3 – 0.5 kg/cm^2 or once in 24 hours whichever is earlier.

d. For dual media filter, gradings must be close & carefully controlled, viz., for sand & anthracite, none of the particle size ratio should exceed 5:1 (anthracite:sand), otherwise terminal velocity of the larger anthracite particles will be the same as or greater than those of the finest sand particles. The specific gravity of anthracite should be 1.4 and that of silica sand 1.6 anthracite size should be between 22 & 12 mesh (0.7 to 1.2 mm) and should be double than that of silica sand (0.45 to 0.6 mm) for effective separation.

e. (1) Filter backwash should be done with filtered water. (2) Backwash rate is generally 24 m/h with air scouring and 36 m/h in absence of air scouring. (3) Treated water quality is less than 2 NTU & 1 NTU in pressure sand filters & gravity filters respectively.

f. The chemical balance of filtered water w.r.to incrustation or corrosion is determined by checking the loss or gain in weights of ferrous coupons exposed to the water within the mains for a period of 90 days, and should not exceed the following:

Level of water above the top of bed	not greater than	0.05 mg/cm^2
Loss of corrosion of galvanised iron	not greater than	5.0 mg/cm^2
Average reduction in alkalinity from inlet to outlet of distribution pipe line	not greater than	1 mg/l

g. Sometimes, there is a loss of residual chlorine from water during its filtration & storage and hence there is practice of having chlorine dose in filter water storage tank also, so as to maintain residual chlorine 0.2 ppm all along in it.

h. Sludge from filter backwash usually contains 0.01 to 0.1% w/v dry solids, while sludge with drain from sedimentation tanks may contain 0.1 to 2% w/v solids.

Chlorination

Chlorination is a supplement in water purification processes after clarification & filtration to disinfect water for specific uses including drinking.

Chlorine kills pathogenic bacteria by the way of interfering with the specific enzymes in the bacterial cells which are vital to support life, but has no effect on spores & certain viruses (e.g. Polio, viral hepatitis) except in high doses. Apart from its germicidal effect, chlorine has several important secondary properties of great value in water treatment. It oxidises iron, manganese & hydrogen sulphide, destroys some sulphites & nitrites, bad taste & odour producing constituents, control algae & slime organism, aids coagulation and hence chlorination is widely used in clarification plant & industrial cooling water. However, caution has to be taken in chlorinating water having too much organics/nitogeneous compounds, which generate obnoxious compounds including trihalomethanes (THM) a very toxic compound. In such an event, chlorination is recommended only after removal of organic/nitrogeneous matters by the way of their sedimentation in clariflocculator & removed by sludge.

Action Of Chlorine In Water

When chlorine is added to water, there is formation of both HCl & HOCl. The HCl is neutralised by the alkalinity of water. The hypochlorous acid ionises to form hydrogen & hypochlorite ions,

$$H_2O + Cl_2 \rightarrow HCl + HOCl$$

$$HOCl \rightarrow H^+ + OCl^-$$

HOCl & OCl- together are called free available chlorine.

The disinfection is mainly due to HOCl and to a small extent due to OCl^-. The best action of chlorine as disinfectant is at pH 7 due to predominance of HOCl at this stage. At pH 8 only, approx. 10% disinfection value remains, because of predominance of hypochlorite ions at this stage. Reactivity of chlorine is also increased with temperature. Chlorination is more effective in water of low alkalinity and low pH values.

However, for proper chlorination it is necessary that water should be free from suspended matter and the amount of chlorine to be added to water should be equal to chlorine demand plus required residual amount after a contact period of 20 - 30 minutes, with a minimum contact period of 10 minutes. The extent of chlorine demand is met by 'break-point' beyond which, if further chlorine is added, free chlorine (OCl + HOCl) only will appear. Suspended solids shield bacteria from chlorine and disinfecting power is reduced due to presence of organic matters also. Effectiveness of chlorine is further reduced due to presence of nitrites, iron, manganese, etc., which react with chlorine.

Subject to the presence of more ammonia or organic nitrogen in water, chlorination can lead to formation of many *halogenated compounds*, some of which are poisonous and carcinogens. Hence, chlorine alternatives are also receiving renewed interest. However, monochloramine & dichloramine (together called as combined available chlorine), if formed with ammonical nitrogen have been marked as loose compounds of chlorine & ammonia, which act as germicide giving better taste of water than chlorine, but have slower action than chlorine. Nitogen trichloride however, if formed is highly poisonous. Phenols, if present, are converted to chlorophenols and free available chlorine is not indicated.

Minimum chlorine dose required for drinking water at 20^0C, which is adoptedunder TPS practices also to chlorinate the filtered water are as follows:

pH value	Minimum concentration of residual chlorine, mg/l (disinfecting period at least 10 minutes)	Minimum concentration of combined chlorine, mg/l (disinfecting period at least 60 minutes)
6 - 7	0.2	1.0
7 - 8	0.2	1.5
8 - 9	0.4	1.8
9 - 10	0.8	Recommended
10 - 11	much greater than 0.8	not recommended

Apart from the reaction strategies of chlorine with water, ammonical/ nitrogeneous/ organic compound, iron, manganese, etc. under pH, temperature & turbidity conditions of water, amount of chlorine dose also depends on nature and concentration of bacteria. The dose in ppm is usually half of the bacteriological oxygen demand in ppm. Since, in gross polluted water with sewage, not only number of bacteria are very large, but also nature of bacteria are very much uncertain, and hence direct chlorination to it should be avoided. Also, the greater the number of bacteria, the longer the time is necessary to reduce them below a given figure. For general disinfection from practical point of view with respect to water borne diseases, free chlorine between 0.1 & 0.4 ppm in a water with pH close to 8 and at 4 ^{0}C, reduction of 99.9% bacteria are obtained in 30 minutes, but when the pH exceeds 8 and below 9, the quantities of free chlorine & contact time can vary by 50%. In a clear water, theoretical contact time of 20 minutes may be suitable. Lower the temperature, higher is the amount of dose.

Under TPS practices, chlorination is also some times done in the raw water reservoir to control organic matters including algae & fungi. Where chlorination has failed to control organic growth, it has been shown that residuals were composed primararily of chloramines or interfering substances had produced false residuals.

Gaseous Chlorine Alternatives

Though there are other oxidising chemicals for microbiological contol, the gaseous chlorine still has been a preferred choice in water treatment practices in respect of technoeconomics particularly when large water are needed to be disinfected. Alternatives are hypochlorite of sodium, calcium, bromine, chlorine dioxide, and ozone, chlorinated or brominated donor molecules like isocyanurales, tri-chloro-s-triazinetriones & hydantoins. Non-oxidising chemicals are also there but they are mostly used in treatment of cooling water.

Chlorine Gas Feed Equipment

The equipment commonly called as 'chlorinator' is designed to maintain the chlorine gas below atmospheric pressure by operating under a vacuum. To prevent the leak, if any, directly pass into the feeding system rather than into the surrounding atmosphere. Since maximum solubility of chlorine at vacuum is about 5000 ppm, the chlorinator is designed to have 3500 ppm only in the feed system, which is then dosed to water stream. A typical flow diagram of a chlorinator is illustrated below.

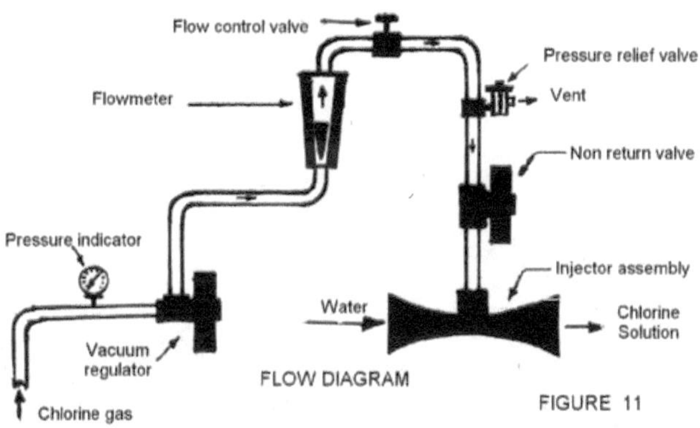

FLOW DIAGRAM

FIGURE 11

For supplying chlorine to the feed system, chlorine gas is normally withdrawn directly from containers (in which chlorine is available in liquid form) when withdrawal rates do not exceed the container vapourisation rate. For large quantity of chlorine gas (viz., for large size water treatment plant & cooling water treatment system), chlorine evaporators are used. A typical flow diagram of an evaporator is shown in figure below.

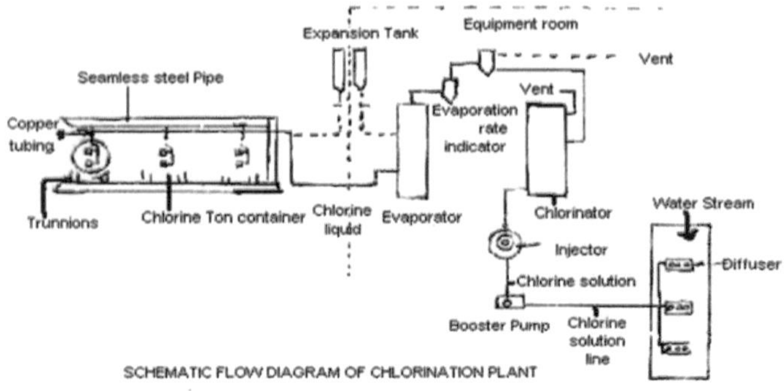

FIGURE 13

The evaporator generally contains a chamber immersed in controlled heated water bath. Chlorine liquid reaching this chamber, maintains the required liquid level (boiling) necessary to meet the vapourisation rate for the gas demand. A change in gas demand will cause the chlorine liquid level to adjust. The released gas then rises through baffles and is superheated as it leaves the chamber to make solution in the chlorinator. A superheated gas is required to prevent chlorine reliquefication in process piping. A full fledged schematic flow diagram of chlorination plant is shown below.

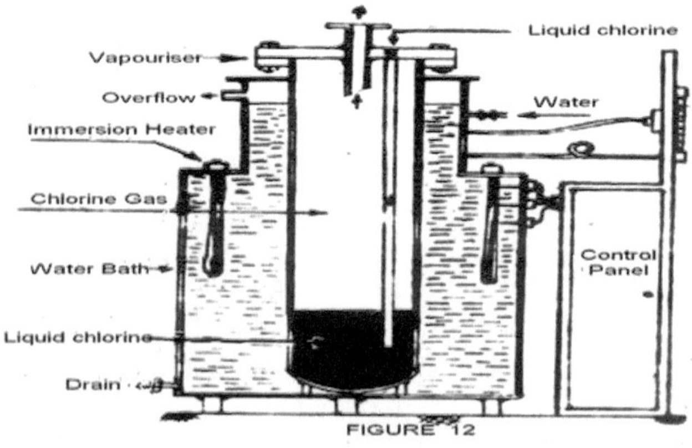

FIGURE 12

Withdrawal of gas from 30/60 kg cylinder should be @1 lb/h and from a ton container @7 kg/h. This will avoid cooling effect to form liquid chlorine to solidify and choke the valve. In case, more amount is required, connect more tubes. Pressure of chlorine should be reduced before regulating valve to 10 lb/h.

It is also desirable that temperature around cylinders be maintained at 20 to 24 ^{0}C and the temperature should not be allowed to fall below 130 ^{0}C.

Post Treatment of Filtered Water/Dechlorination

In water treatment sequence, dechlorination is required for public & industrial water supplies for specific purposes, sometimes partly or even fully. No chlorine residuals again are required for using water for demineralisation of water purposes either by ion exchange resins, membrane processes, or by electodialysis reversal, and similarly in food & beverages processing. Excess free residual chlorine (greater than 0.2 ppm) is also harmful for human consumption. Hence, the chlorinated water as per its further use needs to be dechlorinated either chemically or physically. Common chemical is sodium bisulphite ($NaHSO_4$) and physically by adsorption on granular activated carbon or by suitable aeration.

Activated Carbon Filter: Free residual chlorine present in feed water to demineralisation plant should be removed as it will oxidise the ion exchange resin. Dechlorination by activateal carbon is a surface phenomenon and hence coconut shell based carbon can be used. As an alternative, we can dose sodium sulphite by means of a metering pump. However, carbon filter is preferred since frequent monitoring is not essential.

Softening

In view of the calcium & magnesium ions present in water (causing hardness in water) being concern for scaling, when used in hot water system, viz., cooling cycle, it is imperative to remove these to a certain degree by suitable methods to avoid deposits in a cooling water system. When removed, the water is called soft water and the process is called 'softening'. This helps prepare water for direct use as in cooling cycle/cooling tower make-up or as a first stage treatment followed by ion exchange (demineralisation process) for boiler make-up or process use. In thermal power stations, where high pressure boilers are generally involved, the softened water is led to evaporate under suitable evaporator, the heat energy of steam being utilised in heating boiler feed water and then the ambient condensed water is demineralised through mixed bed ion exchanger.

There are mainly two ways of softening - (1) Lime-soda softening (2) Base exchanger involving strong acid cation ion exchange resins in sodium form.

Lime Soda Softening: In this process, Ca $(OH)_2$ (slaked lime), and Na_2CO_3 (soda or soda ash) are added to hard water. The calcium hydroxide removes the temporary hardness by converting them to an insoluble sludge. Any free carbon dioxide that is present in water will also convert excess calcium hydroxide to calcium carbonate sludge. The calcium hydroxide also converts the permanent magnesium hardness salts to the corresponding calcium hardness salts. Sodium carbonate is then added in the process to remove the permanent calcium hardness salts as below:

$$Ca(OH)_2 + Ca(HCO_3)_2 \rightarrow 2CaCO_3 + H_2O$$
$$Ca(OH)_2 + Mg(HCO_3)_2 \rightarrow 2CaCO_3 + Mg(OH)_2 + H_2O$$
$$Ca(OH)_2 + CO_2 \rightarrow CaCO_3 + H_2O$$
$$Ca(OH)_2 + MgSO_4 \rightarrow Mg(OH)_2 + CaSO_4$$
$$Ca(OH)_2 + MgCl_2 \rightarrow Mg(OH)_2 + CaCl_2$$
$$NaCO_3 + CaSO_4 \rightarrow CaCO_3 + Na_2SO_4$$

Sodium carbonate is then added in the process to remove the permanent hardness salts with following reactions:

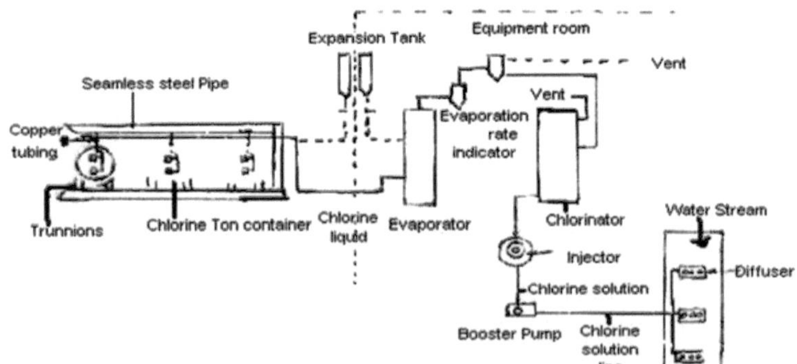

SCHEMATIC FLOW DIAGRAM OF CHLORINATION PLANT

FIGURE 13

Nowadays, hot lime softening is not used in the industry because of its complexity of process and operational problems associated with it. Only cold lime softening is used.

Silica reduction: Hot process softening can also provide very good silica reduction, which is accomplished by adsorption of silica on the magnesium hydroxide precipitate. If there is insufficient magnesium present in the raw water to reduce silica to the desired level, magnesium compounds (such as magnesium oxide, magnesium sulphate, magnesium carbonate or dolomotic lime) may be used.

Dealkalisation: Treatment by lime precipitation reduces alkalinity. If raw water alkalinity exceeds the total hardness, sodium bicarbonate alkalinity is present. In such cases, it is usually necessary to reduce treated water alkalinity in order to reduce condensate cooling system corrosion or permit increased cycle of concentration. Treatment by lime converts the sodium bicarbonate in water to sodium carbonate as below:

$$NaHCO_3 + Ca(OH)_2 \rightarrow CaCO_3 + Na_2CO_3 + H_2O$$

Calcium sulphate (gypsum) may be added to reduce the carbonate to required levels, with the following reaction:

$$Na_2CO_3 + CaSO_4 \rightarrow CaCO_3 + Na_2SO_4$$

Defluoridisation: Fluoride in raw water if present in excess of limits described in drinking water standard, needs to be removed. Many processes are available for removal of fluoride. Reverse osmosis is one of the suitable processes. Further treated water from RO can be remineralised for drinking purpose.

Reduction of other contaminants: Lime softening processes with suitable filters are able to reduce oxidised iron and manganese, organics (colour contributing colloids), turbidity (with use of cationic polymer in case of high turbid water before the water enters the softener vessel) and oil by adsorption on the precipitates formed during treatment. However, higher oil concentration should be reduced before lime treatment; otherwise oil could have dispersing influence to cause floc carryover.

Base Exchanger: This method of softening is most widely applied in industries including thermal power stations, which involves passing the water through a vessel containing a bed of strong acid cation exchange resin in sodium form. The calcium and magnesium ions of the water are exchanged for sodium ions.

$$RNa^{+} + CaSO_4 \rightarrow RCa^{+} + Na_2SO_4$$

An exhausted resin is then converted again into its original state by regenerating same with a 10% solution of sodium chloride charge:

$$RCa^{+} + NaCl \rightarrow RNa^{+} + CaCl_2$$

Through base exchanger softening process, hardness level of raw water can be brought down to 5 ppm in the treated water, and the operation is simple and reliable. However, the total dissolved solids content, alkalinity and silica in the water remain unaffected, but has a very low scaling and corrosive tendency, if any, can be avoided by suitable blending of the soft water with the original hard water. Alkalinity in the treated water is now in the form of sodium bicarbonate.

When the operating conditions require the removal of hardness and alkalinity both without removal of other solids, viz., for make-up to a cooling system so as to eliminate troubles of precipitating out of magnesium hardness as $Mg(OH)_2$, then a process consisting of a weakly acidic cation exchanger (to remove Ca^{++} & Mg^{++} ions attached with bicarbonates and carbonates) followed by a mechanical scrubber tower (to remove CO_2) and base exchange unit (to remove Ca^{++} & Mg^{++} ions attached with chloride, sulphates, etc.) are employed.

$$[Ca/Mg/2Na].\,2HCO_3 + 2\,R.H \rightarrow 2R[Ca/Mg/2Na] + 2H_2CO_3$$

$$[Ca/Mg].Cl/SO_4/NO_3 + 2\,R.Na \rightarrow R[Ca/Mg] + 2NaCl + Na_2SO_4 + NaNO_3$$

Thus, ion exchange dealkalisation systems produce hardness free, low alkalinity water at a reasonable cost and with a high degree of reliability. Apart from cooling cycle, they are also suited for processing feed water for low pressure boilers. The above combination of treatment also reduces the total dissolved salts.

There are other methods of dealkalisation also including direct neutralisation with acid, base exchanger/strong acid cation exchange and base exchanger/chloride anion exchange, but the above combination with weak acid cation exchange is most acceptable under TPS practice.

Demineralisation

After dechlorination, the next stage of treatment is demineralisation by Ion Exchange process, by a suitable DM plant

The DM plant conventionaly consists of:

- Strong Acid Ction exchanger (SAC)
- Degasser (DG)
- Strong Base Anion exchanger (SBA)
- Mixed bed (MB) polisher
- Ultrafiltration (UF) unit

SAC converts all the salts present in feed water into acids. Alkaline salts are converted into carbonic acids (H_2CO_3) while the neutral salts are converted into mineral acids (HCl, H_2SO_4, HNO_3). The unit contains a strong acid cation exchange resin in hydrogen form.

On exhaustion, the ion exchange resin is regenerated either by HCl or by H_2SO_4 Regeneration is carried out either in co-flow or counter current (CCR) mode. CCRmode is preferred as sodium slip from the unit can be reduced to less than 1 mg/l.

Carbonic acid is a weak acid and dissociates into H_2O and CO_2. Degasser removes the carbon dioxide by blowing air through a column packed with rasching rings. SBA neutralizes the mineral acids present in the effluent from the cation unit. It also removes the residual CO from the degasser and weakly ionised silica. The only impurity now present in treated water from SBA is NaOH present due to sodium slip

The SBA unit contains a strong base anion exchange resin in hydroxide form. On exhaustion the unit is regenerated by NaOH. Regeneration is carried out either in Co flow or Counter current (CCR) mode. CCR mode is preferred as silica slip from the unit can be reduced to less than 0.2 mg/l as SiO_2.

MB unit polishes the treated water from the SBA unit. And produces high purity treated water with conductivity less than 2 µsiemen/cm and silica less than 20 ppb. The unit contains a mixture of a strong acid cation exchanger and a strong base anion exchanger On exhaustion, the resin is separated by backwashing and regenerated separately

UF unit removes colloidal silica present in water. Treated water is now highly pure and is stored in the DM water tank

Removal of colloidal silica results in reduced blowdown which in turn results in savings in fuel, water and chemicals

Polystyrene—Divinyl Benzene resins are used in majority of ion exchange applications. The resins which have ionic sites consisting of mobile charge of 'SO_3H' radicals & mobile sodium cations (Na^+) remove the cations present in water (hence called cation exchange resins) and the resins which have tertiary or quaternary ammonium group as mobile cationic radicals & mobile chloride anions (Cl^-) remove the anions including silicic and carbonic acids present in water (hence called anion exchange resins).

Synthetic ion exchange resins of polystyrene group are generally in use for deionisation purposes. For cation removal (viz., Ca, Mg, Na, etc.) strongly acidic cation exchange resins as below are used:

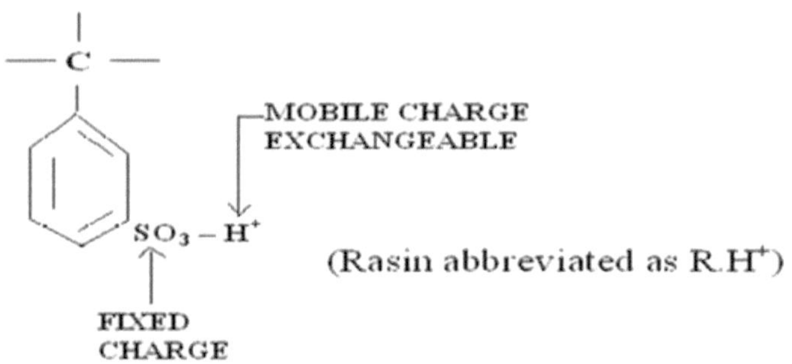

For anion removal (e.g. Cl, SO_4, NO_3, etc.) strongly basic anions exchange resins as below are used:

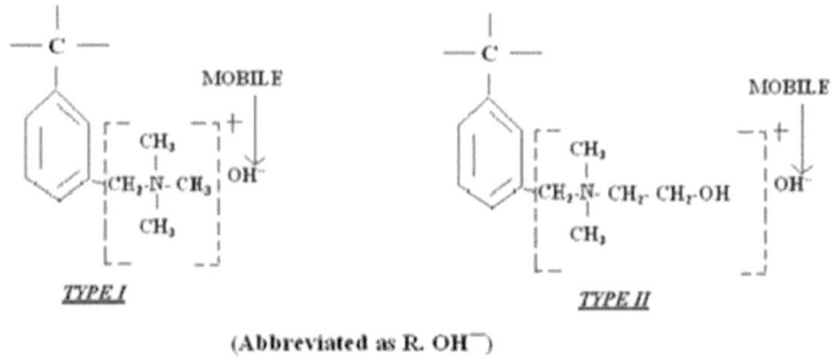

Exchange Reactions

$$RH^{+} + Ca(HCO_3)_2 \rightarrow R\,[Ca/Mg/Na] + H_2CO_3/HCl/H_2SO_4$$

$$ROH^{-} + H_2CO_3/HCl/H_2SO_4 \rightarrow R\,[H_2CO_3/HCl/H_2SO_4] + H_2O/\,H_2O/\,H_2O$$

After the exchange reactions as above, when resins are exhausted, they are recharged to work as before, by reactions with (a) dilute acids preferably HCl, in case of cation exchange resins and (b) dilute alkalis preferably NaOH, in case of anion exchange resins, as below:

Recharging Resins

$$R\,[Ca/Mg/Na] + H_2CO_3/HCl/H_2SO_4 \rightarrow RH^{+} + CaCl_2/MgCl_2/NaCl$$

$$R\,[H_2CO_3/HCl/H_2SO_4] + NaOH \rightarrow ROH^{-} + NaHCO_3/NaCl/Na_2SO_4$$

Similar type of reactions occur when resins are used for softening purposes (viz., removable of hardness causing cations like Ca & Mg) and not for total deionisation. For softening, same strong acid caton resins are used, but in the sodium form of the resins, described as R. Na^{+}.

During exchange reaction, it behaves like:

$$RNa + Ca/Mg(HCO_3)_2 \rightarrow R\,[Ca/Mg] + Na_2CO_3$$

& during recharging, when dilute solution of NaCl is used (usually 10%)

$$R(Mg/Ca) + NaCl \rightarrow R\,Na + CaCl_2/MgCl_2$$

Ion Exchange Resins:

Introduction:

Ion exchange resins are synthetic polymers that are insoluble in all solvents. They are capable of reacting like acids, bases or salts. The resins, however, differ from acids, bases and salts in one fashion. Only the cations (in cation exchange resins) or anions (in anion exchange resins) are free to take part in chemical reactions. Those exchangers in which the anionic portions are able to react are called anion exchangers whereas, the ones in which the cationic portions are able to react are called cation exchangers. In aqueous media and sometimes in non-aqueous media, cation exchange resins are able to exchange their cations with other cations and similarly anion exchange resins are able to exchange their anions with other anions.

The basic polymeric structure or solid support for ion exchange resins is called the matrix. The matrix is a three dimensional co-polymer on which acidic or basic sites are situated. These acidic or basic sites are called functional groups or functionality of the ion exchange resin. The matrix along with these functional groups is called a functional polymer or ion exchanger or simply resin.

The mechanism of ion exchange is reversible and hence ion exchange resins can be repeatedly converted from one ionic form to another and reused. The process of converting the ion exchange resins to the desired ionic form at the end of cycle of its intended use is known as 'regeneration'.

Ion Exchange resins can be broadly classified into following four categories. Strong Acid Cation exchange resin and Weak Acid Cation exchange resin. Strong Base Anion exchange resin and Weak Base Anioin exchange resin

Strong Acid Cation Exchange Resin:

All the major strong acid cation exchange resins involved in industrial water treatment applications have a chemical matrix consisting of styrene and divinylbenzene. The functional groups are sulphonic acid radicals. These resins differ mainly in divinylbenzene content or matrix structure (gel/isoporous/macroporous).

Generally, in water softening, a strong acid cation exchange resin with 6 - 8% divinylbenzene crosslinking is used. INDION 220 Na and INDION 225 Na are the examples of this resin. For demineralisation the main resin that is used is the cation resin in hydrogen form ie., INDION 225 H.

Weak Acid Cation Exchange Resin:

In industrial water treatment applications, weak acid cation exchange resin is used primarily where there is a high degree of hardness and a high degree of alkalinity. This resin has the capacity of exchanging all cations associated with alkalinity to a much greater degree than strong acid cation exchange resins. A weak acid cation exchange resin consists of polyacyrlic acid divinylbenzene matrix with carboxylic functionality and gel structure. The major advantage of this resin is that it can be regenerated with stoichiometric amounts of regenerant, and is therefore, much more efficient. INDION 236 is an example of this resin.

Strong Base Anion Exchange Resin:

Strong base anion exchange resins can be divided into two categories called Type I and Type II. The Type I resin has the highest overall basicity, and, therefore, gives best effluent quality. The Type II resin also removes anionic constitutents, but has as lower basicity, and therefore, requires less caustic during the regeneration cycle. In general, a Type II strong base anion exchange resin is recommended where silica effluent quality is not as critical, and also where a relatively high chloride and/ or sulphate content prevail in the raw water. INDION FFIP is an example of strong base Type I and INDION NIP is an example of strong base Type II anion resin.

Weak Base Anion Exchange Resin:

In the field of deionisation, weak base anion exchange resin is used primarily to remove strong acids such as hydrochloric and sulphuric acids. Unlike strong base anion exchange resin, weak base resin does not have the capability to remove bicarbonate and silica. But it has a much higher capacity for the removal of chlorides and sulphates. The resin can be of gel or macroporous structure. For treatment of water which does not present organic fouling problems, the gel type weak base resin is used. For treatment of water containing organic contaminants (humic and fulvic acids), macroporous weak base anion resins are preferred. INDION 850 is an example of macroporous weak base anion resin.

Characteristics of Resin:

The characteristics of ion exchange resins can be broadly classified into three categories.

1. Physical characteristics
2. Chemical characteristics
3. Operating characteristics

Physical Characteristics:

Particle Size Distribution:

Particle size distribution is a very important physical characteristic of resins. Many factors such as rate of exchange (kinetics), backwash expansion, pressure drop across the bed and susceptibility to breakage are affected by particle size of the resin.

Thus, on one hand we have coarse particles which show:

- Poorer rate of exchange
- More attrition losses due to the fact that larger particles, undergo higher volume change during operation and hence, are more prone to develop cracks
- Lower pressure drop across the column

On the other hand, we have fine particles which give:

- Good backwash expansion
- Better rate of exchange
- Higher pressure drop

Higher pressure drop is undesirable since it exerts greater force on resin bead, thereby breaking it physically and generating more fines. Thus, it is a vicious circle of fines leading to high pressure drop which in turn generates additional fines.

In view of all these problems associated with coarse particles as well as with fine particles, it is necessary to have a resin which is neither very coarse nor very fine. Ion exchange resins used for conventional water treatment application are of screen grading 14 BSS to 52 BSS i.e. between 1.2 mm to 0.3 mm.

Particle size distribution of ion exchange resins is determined by carrying out wet sieve analysis of a representative sample of resin and measuring percentage of resin retained on each sieve. From this, a plot of cumulative percentage retained on screen opening is drawn on a log-log scale. From the plot screen opening corresponding to 90% retained and the screen opening corresponding to 40% retained is noted.

Effective size is defined as the screen opening that retains 90% of the sample. The ratio of screen opening corresponding to 40% retained and 90% retained is defined as uniformity coefficient. Particles smaller than 52 BSS (0.3 mm) are defined as 'fines'. Typically, ion exchange resins have effective size between 0.40 to 0.50 mm, uniformity coefficient below 1.7 and fines content below 1.0%.

Cracks & Pieces:

Ion exchange resins (during operation) are continuously subjected to osmotic and physical shocks. The stress and strain resulting from these shocks may lead to cracks developing in certain resin particles. Cracked beads gradually turn into pieces. These in turn lead to high pressure drop across the bed and hence are undesirable.

Cracks and pieces in resin samples are determined by microscopic examination of a representative sample.

Density:

Anion exchange resins have typically a true density of 1.05 to 1.15 g/ml while cation exchange resins are heavier, having true density of around 1.20 to 1.30 g/ml. Adequate density difference between cation and anion resins is necessary for their effective separation in mixed bed units.

The true density of resins is important in also assessing the backwash/settling characteristics and maximum concentration of regenerant that can be used. Apparent density (bulk density) of ion exchange resins helps to know the shipping weight. From the figures of true and apparent densities, void volume can be computed.

Surface Area:

Ion exchange resins are spherical in shape. Gel type or isoporous resins have negligible total surface area of less than 0.01 m^2/g. This is because these types of resins have only the external surface area. On the other hand macroporous resins have very large surface area (atleast 5 m^2/g and often around 30 - 40 m^2/g) due to presence of large number of macropores within each bead. Measurement of surface area indicates the extent of macroporosity in the resin. This is carried out using a BET instrument.

Colour - Throw:

As mentioned earlier, ion exchange resins have cross linked polystyrene matrix. The crosslinking is introduced to ensure that they are insoluble in water or any other solvent. During the manufacturing process, however, a very small amount of linear polymer is also formed. During subsequent sulphonation for manufacture of cation exchanger, this linear polymer gets converted to polystyrene sulphonic acid, which is water soluble. This is what gives colour-throw from cation exchange resins during first few cycles. Colour-throw is undesirable not only because it shows inferior quality of cation resin, but also because it has an adverse effect on the down stream anion exchange bed.

Chemical Characteristics:

Water Regain:

Water regain is defined as the amount of water chemically bound with one gram of dry resin. It indicates the extent of crosslinking in the resin. Higher the water regain, lower is the crosslinking. Lower the water regain, higher is the crosslinking in the resin.

Ion exchange resins with low water regain (high crosslinking) are characterised by:

- Poor kinetics and low operating capacity
- Higher ionic leakages
- Poor elasticity and hence brittleness

On the other hand, resins with high water regain (low crosslinking) are characterised by:

- Better kinetics
- Relatively soft and of lower physical strength

Typically, resins manufactured by us for conventional water treatment applications have water regain from 0.90 to 1.20 g/g. Increase in water regain with use indicates decrosslinking taking place due to presence of oxidising agents.

Total Exchange Capacity (TEC):

Total exchange capacity of a resin indicates the total number of exchange sites available in the resin. This is a function of the manufacturing process. Total exchange capacity of resins is expressed in several ways. For example, it can be expressed as milliequivalents of exchangeable ions available per dry gram of resin. It can also be expressed as meq/wet g or meq/ml of resin. In water treatment plant operations, the total exchange capacity is often expressed as g $CaCO_3$/l of resin.

Sodium Chloride Value:

Sodium chloride value or salt splitting capacity of a cation exchange resin gives the number of milliequivalents of exchangeable hydrogen ions in the resin that are sufficiently acidic to split neutral salts.

Strong Base Capacity (SBC):

Strong base capacity of an anion exchange resin is a measure of the strong base (quaternary ammonium) groups in it. A strong base Type I or Type II resin like INDION FFIP or INDION NIP has more than 95% of its capacity as the strong base capacity.

The strong base capacity drops with increased use of resins. This drop is much more pronounced in case of Type II resins. Drop in SBC affects the ability of resins for silica removal. However, as in case of TEC, only a significant drop in SBC (say by 40 to 50%) can truly affect the ability of resin to remove silica.

Operating Characteristics:

Operating Capacity:

Operating capacity of an ion exchange resin is the capacity obtained under actual plant operating conditions. It is substantially lower than the total capacity of the resin, because only economic quantity of regenerants are used in the plant.

Contrary to the general misbelief, operating capacity is not proportional to the total capacity of the resin, but a function of several characteristics and operating parametres such as:

- Regeneration level
- Temperature
- Regeneration time
- Exhaustion time (Flow rate)
- Ionic ratios in inlet water quality
- End point ionic leakages
- Bed depth
- Cross linking resin

The method of determining operating capacity consists of repeated cycles of backwash, injection, rinse and exhaustion of the sample in a 1" diametre column. Generally, a 200 ml sample is used for this evaluation and feed water of known composition is used for exhausting the resin. The cycle is stopped when a predetermined level of leakage is observed. The results are expressed as ions removed from inlet water (expressed as g $CaCO_3$ /l of resin), under given operating conditions.

Resin Fouling:

Ion exchange resin can become fouled with contaminants that hinder the exchange process and affect the useful resin life. Some of the common foulants present in water and their effect on resin performance are discussed below:

Iron/Heavy Metal Fouling:

Iron is deposited on the ion exchange resin beads in the form of hydroxides or trivalent oxides. Iron deposits on the resin not only blocks the pores but also catalyses oxidising action of the dissolved oxygen in water.

Iron fouling of cation resin is noticeably less when hydrochloric acid is the regenerant. In case of softeners or units which are regenerated with sulphuric acid, occasional wash with dilute hydrochloric acid removes most of the iron deposits on the resin. Iron fouling of the anion resin can take place due to presence of iron impurity in regenerant solution.

Excessive quantities of aluminium sulphates used in pretreatment plant in some cases cause fouling of the cation resin with aluminium in the same manner as iron. Aged aluminium oxide deposit on resins is difficult to remove. The most effective method of preventing fouling by heavy metals is to eliminate these elements at source, rather than doing cleaning operations on the fouled resin bed.

Organic Fouling:

Large molecular weight organic acids like fulvic and humic acids, which are formed by decomposition of vegetable matter in water, can get adsorbed on anion exchange resins but do not get eluted during regeneration. An organically fouled resin results in lower operating capacity and high rinse requirements.

Most organic matter is colloidal in nature and therefore can be removed during pretreatment stage by coagulation and filtration. Use of activated carbon filters is also known to help in reduction in the content of organic matter in water.

The resin structure also plays an important part in determining the fouling resistance of the resin. Isoporous resins, which have equal porosity so that entanglement of large molecules with resin is eliminated, have been found to be effective. Macroporous resins have large macropores which allow large molecules to freely enter and leave the structure.

Use of scavenger resin beds (which are macroporous) reduces the fouling of the down stream strong base resins to a considerable extent. Some improvement in performance of organically fouled resins can be obtained by alkaline brine wash of the resin.

Oil Fouling:

Presence of oil in water can cause physical fouling of the resin surface. Oil fouling can be prevented by good pretreatment. Oil fouled resin can be cleaned by washing with caustic soda solution or with suitable surface active agent.

Polyelectrolyte Fouling:

Use of polyelectrolytes in pretreatment of water must be properly controlled. Excessive quantity of polyelectrolyte (which are high molecular weight organic substances) can cause irreversible fouling of the cation/anion resins depending on the type of polyelectrolyte involved.

Precipitation:

If a high concentration of sulphuric acid is used during cation regeneration, calcium sulphate precipitation can occur. Precipitation of calcium sulphate will cause problems such as high pressure drop and leakage of hardness ions during service, and some times break the bead. Calcium sulphate precipitation can be cured by backwash or by a hydrochloric acid wash.

Resin Storage:

The resin is despatched in moist condition, packed in either drums or bags. On receipt, keep the resin drum upright in shade to prevent the same from drying. If, for some reason, the drums have to be left in the sun, cover the drums with tarpaulin and keep the tarpaulin wet with water. Ensure that the drums are shifted to a sheltered area as soon as possible.

At intervals of 2 - 3 months, check the resin drums in your store. If the resin has lost some wetness, add some demineralised water to moisten the same. When the resin is charged into the process, ensure once again that the resin is always kept under water. Do not keep the vessel drained for long, as this may cause the resin to dry up, especially in plants situated in the open and exposed to direct sunlight. If the plant is to be shutdown for a long period of time, keep the resin under water and in exhausted condition. It is even beneficial to soak the resin in a dilute (approximately 10%) solution of brine during these long shutdowns.

In case, the resin is required to be removed from the vessel due to maintenance work on the unit, ensure that the resin removed is stored well and kept moist all the while, until it is ready to be recharged into the vessel. The resin is important item of the plant and a costly one too. It is essential then, that it is well stored and cared for.

Resin Maintenance:

In order to always obtain peak water quality, it is essential to maintain the resins in the right condition. This calls for some special care. The ion exchange resins are bound to undergo some changes with usage. The changes sometimes are natural to the resin; but mostly the deterioration of the resin is due to factors such as poor influent water quality or inadequate maintenance. One would, no doubt, like to see the resin working efficiently for a sufficiently long period and for this reason, it is necessary to carry out some regular maintenance on the resin. Some of the items that should be taken care of are given below:

Check first the nature of water that enters the demineralising plant after pretreatment and filtration. This water should have good clarity; all foulants like heavy metals and organic pollutants should be completely excluded in the feed water to the exchanger. The pretreated water should be oil free and any chlorine leftover from upstream chlorinating system must be removed.

Turbid feed water can clog resins while increasing bed pressure. Heavy metals, organics and oil can foul the resin, reducing its capacity and affect treated water quality also. Free chlorine can cause irreversible decrosslinking of the resin. What is true for the feed water equally hold for the regenerant that are used. Use the best quality regenerants for optimum results and enhanced life of the resins. Check every batch of regenerant and see that they conform to the relevant IS specifications.

Even though utmost care is taken to minimise the impurities at the source of water or the regenerants, there can be a slippage of these down through the DM plant. In such cases, few treatments become necessary to restore the resin back to its near original state. If such maintenance is not done, it could result in poor treated water quality, reduced plant capacities and higher expenditure, for regeneration of the resins.

A resin has become te dirty can be cleaned by a thorough backwash given for an extended period. Open manhole backwashes given regularly also help. Maintain the resin bed in a clean condition. Air scouring of the bed is a further aid to cleaning and helps improve resin condition considerably when backwash alone does not help.

- The resin iscontaminated with heavy metals, an overnight soak in 10% HCl would help decontaminate the resin and provide better capacity and treated water quality
- Resins that are reversibly fouled by organics can be safely cleaned up with an alkaline brine solution. This treatment is better when the solution is poured for a longer period over the resin. In most cases an overnight soaking is adequate. Hypochlorite solutions are known to be more effective organic defoulants than alkaline brine, but at the same time, cause some damage to the resin due to the oxidising nature of the solution
- Oil fouled resins are amenable to cleaning with a dilute solution of INDION 2523
- Silica fouled anion resins are best rejuvenated with a hot (50° C) solution of caustic soda. Check the type of anion resin in the unit. If the unit contains a Type II strong base anion exchanger, do not employ hot caustic soda, instead carry out double alkali regeneration employing twice the normal quantity of alkali

The frequency of clean up will vary on a case to case basis. While an annual, six-monthly or quarterly treatment can be considered a normal schedule, a too frequent requirement, for cleaning of resins would indicate that the inputs to the exchanger need to be thoroughly investigated.

In the ideal case of excellent feed water - clear and free of contaminants - the resin will work to give long and trouble free service. There is, of course the natural decay that takes place in every resin and about which one can do very little. This decay varies from one resin type to another and it takes quite some time before natural degradation begins to interfere with plant water quality and treatment economics.

Besides the chemical ageing and fouling, the resin is susceptible to physical breakdown into finer particles. This is caused by a number of factors such as pressure drop, osmotic shock, thermal shock and volume changes taking place with every regeneration and exhaustion cycle. The formation of resin fines is, therefore, a normal phenomenon and the removal of these fines in good time is recommended.

Observe the unit pressure drops. These should remain constant at a particular flow rate. As the drop across the bed increases, the resin bed should be given a backwash at a higher flow rate and for a longer period than normally given. Fine resins should be washed out to drain. It is beneficial to give open manhole extended backwash for removal of the fine resins. After the backwash, scrape off the top few centimetres of the bed.

Whenever the unit is opened, use this opportunity to check the unit's internal condition. Resin can get contaminated due to the corrosion of the vessel and its internals.

Do not forget to check resin bed depths whenever the unit is opened for backwash or internal inspection. Much of the capacity drop problems may be attributed to a lower volume of resin in the unit. Have the resin sample from the unit analysed and after careful consultation - top up any shortfall or replace the resin.

Resin Sampling:

A resin may be sampled for on-the-spot checks or for a detailed examination

The sample may be drawn from drums in the storage yard or from the units in the water treatment plant. The sample that is collected is important and should be representative of the bulk. Hence, it is necessary to follow some standard procedures while sampling.

For a resin charged in a unit, it is advisable to collect the sample after the resin bed is exhausted and has been backwashed and allowed to settle. In case of mixed bed, it is preferable to take three samples - composite, cation and anion resin.

If samples are required at specific levels they may be collected from say, top, bottom or at every foot of resin depth as desired. When not explicitly mentioned, it is a normal practice to collect a sample approximately one foot below the top surface of the bed to represent the resin in the unit.

A more representative sample from the unit, however, would be to sample the resin entirely along its depth with the help of a plastic tube. Where this is difficult to accomplish the 'one foot below top surface' approach could serve as the standard resin sample. Resins are charged into a mixed bed, samples can be drawn from individual cation and anion layers as suggested above. Alternatively, it is a good practice to collect the sample while the resins are being air mixed. The sample collected provides a good idea of the effectiveness of mixing and resin ratios. Similarly in the case of layered beds, the weak and the strong function resins can be sampled by collecting resin from the individual layers. Just as in the mixed bed unit, it is going to be difficult to get the 'one foot' sample from the bottom resin layer. In such a case, take advantage of the resin outlet pad provided for getting a representative sample.

A sample may be collected for checking, when in doubt, about resin performance but it is good engineering practice to sample a resin at fixed intervals. Annual sampling and checking of resin condition should be adequate.

More important than regular sampling is the initial sample that one should collect from every unit at the time of charging the resin. This control sample averages, so to say, the various resin batches charged into the unit. It also serves as a 'control' sample for comparison of all future samples from the unit.

One may want to check the condition of resins in the store, especially when they have been stored for a very long time or when there has been some mix up in the storage or difficulty in identifying the type of resin. In such a case, send samples from each drum or combined samples depending on the problem at hand. While sampling a resin from each drum collect samples at different levels and make a composite sample for each drum. The quantity of resin required will depend on the nature of tests required. In most cases, a 500g sample would be adequate for all tests.

Simple but an important thing to remember after resin sampling is to label the container, giving the name of the organisation, the contact person and some details of the resin, namely, the unit from where it has been drawn, the nature of the resin sample i.e. exhausted or regenerated, and the date of sampling. A bit of resin history would also help in advising further course of action.

Ion Exchange Vessels

Cation, anion & mixed bed exchanger vessels with resin regeneration facilities are shown in Figure 14 to 19 respectively. All exchange vessels, usually vertical steel (rubber lined) are pressure vessels with dished ends and have system for bringing up to service and regenerating the resin with necessary control valves & piping, inlet distribution system, free board space, regenerant distribution system, ion exchange resin bed and a resin retaining underdrain collecting system.

All exchanger vessels operate through two basic cycles:

a. The service cycle (it produces deionised water).

b. The regeneration cycle (it restores resin capacity after exhaustion).

The regeneration cycle however is mostly accompanied by backwash cycle of resins.

In the service cycle, water enters the exchanger vessel either through bell mouth assembly or inlet distribution system and flows through the resin bed, which deionises the water. The deionised water is then collected in underdrain system & passed on for further processing or stored for onward use. The required amount of resins must rest on strainers, below which lies underdrain collector system. The exchange reaction i. e. deionisation reaction kinetics are most efficient, when resin bed is in fully compact form and service flow rate through it is kept between 15 m/h and 25 m/h, which is decreased or increased as per total dissolved salts loading of inlet water i. e. higher the dissolved salts lower should be the inlet velocity. At very low flow rates, the water is however not sufficiently distributed and optimum resin- water contact cannot take place.

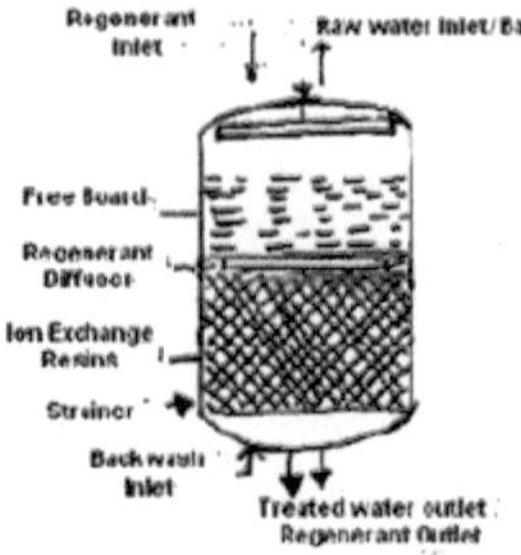

COCURRENT REGENERATION

FIGURE 16

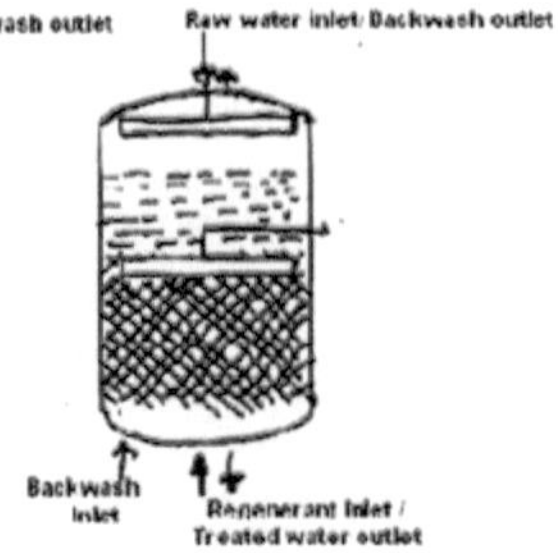

COUNTERCURRENT REGENERATION

FIGURE 17

When the resin is exhausted, the regeneration of resins is accompanied generally after backwash. This is required in view of resins getting dirty with suspended materials present in the inlet raw water. The backwashing not only removes accumulated material but also reclassifies the resin bed, for which due free board space is being made in the exchanger vessel usually 50 to 75% of the resin bed volume. In the backwash step, water flows from underdrain collector system through the resin bed and out of inlet distributor to waste. Backwash flow rates usually vary between 10 to 20 m/h per m^2 of resin bed. After the backwash is complete, due concentration of regenerant is introduced to the compact bed of resin through regenerant distributor and conventionally flows down through the resin bed at a slow rate usually between 1.5 m/h and 3 m/h per m^2 of resin bed. Regenerant waste is collected through underdrain & sent to waste. To achieve optimum efficiency of regeneration, regenerant concentration should be 4 - 5% in general (Max 2.5% in case of H_2SO_4) in case more calcium alkalinity is apprehended during cation resin regeneration, rather with more calcium/magnesium it is better to have two stages regeneration, initially with 1 to 1.5% and then with 2.5% sulphuric acid. In case of weak acid/weak base exchange resins 2% acid/base in the form of regenerant is used, and when the cation resins are used as base exchanger i.e. softening, 10% regenerant in the form of NaCl is used.

Supersaturation of $CaSO_4$ has to be avoided during regeneration of cation resins with H_2SO_4 (which will foul the resin) by regenerating it at or below 25^0C, flow 3.5 $m^3/h/m^2$ and concentration below 5%, rather stage wise starting with much lower strength of sulphuric acid.

As per techno economics, waste water generated during regeneration should be limited to an optimum quantity. For cation resin bed, regeneration on sulphuric acid, the waste water generation should not be more than 11 BV of resin and on HCl regeneration the same should be limited to 10 BV. For anion bed regeneration on NaOH the same should not exceed 12 BV of resin.

The conventional method of regenerating resin called as coflow regeneration, is however economic for water having low ppm of dissolved salts particularly when effluent leak, viz., sodium from cation exchanger and silica from anion exchanger are not of extringent standard.

Hence, to have efficient regeneration dealing with slightly abnormal water during service, counter current regeneration or thoroughfare regeneration techniques are employed. The counter current regeneration is also called upflow regeneration in which middle collector is provided, lying just below the upper level of the resin bed and fitted with strainer nozzles or mesh covered holes through which resin cannot pass. These units however run in normal downflow service. During regeneration, the regenerant of due concentration is pumped into the bottom distributor to pass upwards through the bed & at the same time air or water enters the top distributor to hold the bed packed & immobile. Both the flows leave the unit through buried middle collector to waste.

The thoroughfare regeneration is generally employed where both weak and strong acid/ base resin beds are used in joint operation. The waste regenerant of strong acid/base resins is utilised to regenerate weak acid/ base resins either coccurently or counter currently.

Mixed Bed Exchanger Vessel

By using successive cation & anion exchangers, though water can be completely demineralised, but it shows that these two steps, if repeated atleast once, a product of high purity can be achieved and more so when the two resins are mixed together intimately. Such mixing facilitates efficient kinetics of reaction in view of cations & anions resin lying always side by side making reversible reaction of exchange more in the forward direction, and thus a product of very highest purity results removal of weak acids like carbonic & silicic acids. In a TPS presence of these weak acids in DM water is not tolerated which is used in steam generation & rotating turbine. A mixed bed exchanger vessels along with its regeneration techniques is shown below.

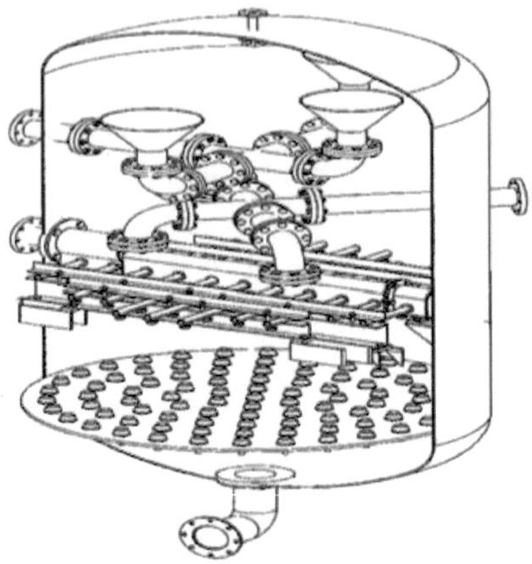

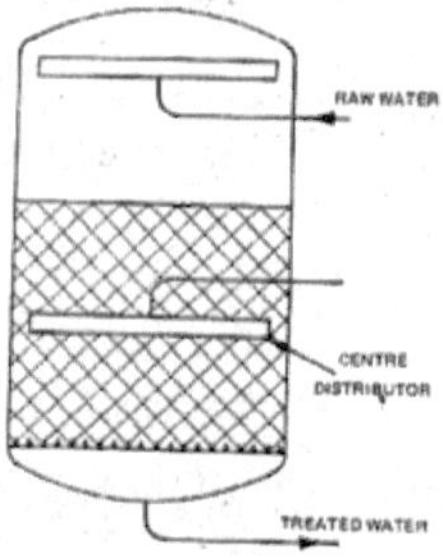

MIXED BED
FIGURE 19

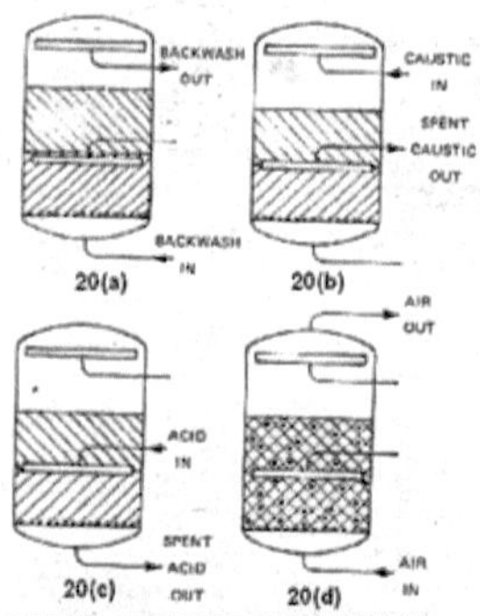

FIGURE 20 MIXED BED REGENERATION

The mixed bed cycle depends on separating the two resins first for regeneration individually, and remixing them for the treatment of water. Anion resin has a lower density than cation resin and a mixture of the two can therefore be separated by upflow classification i.e. by backwashing. After separation, first the anion resin, which settles above the cation resin bed, is regenerated usually by passing alkali solution from top of anion resin bed, but taking the waste alkali out from the centre distributor, so that alkali does not contact the bed of cation resin. To ease this operation, an upflow of water of lower value is also given from bottom of the vessel to pass through the cation bed.

After completion of regeneration of anion resins, cation resins are regenerated with the same precaution of not allowing any contact of acid regenerant with the anion resin, and for this purpose, some downflow of water through anion bed is given. When both the resins are duly regenerated, the water above the total resin bed is drained up to few inches above the surface of the bed, after which air is blown into the vessel through the bottom distributor with release of air from top of the vessel. This mixes the resins thoroughly after which the bed is pressurised by filling it with water immediately so that resins are stopped separating again. Then a final rinse is given to the total bed before being returned to service.

In view of the mixed bed treating almost pure water, a high water velocities are employed, usually 35 - 55 m/h, which reduces the size of the mixed bed & its overall capital cost. Actual velocities go as high as 120 m/h in condensate polishing (which is also a mixed bed). However, in these cases not only centre distributor is made strong enough to withstand the large pressure losses and the resulting force, it is a normal practice to build a separate regeneration plant, where resins from the mixed bed are transported for regeneration and after regeneration & mixing is retransported to the mixed bed for final rinsing & putting into service.

The minimum bed depth of resin in a mixed bed should be 500 mm or flow x 0.025 m^3 volume. Resin particle size for this bed have been upgraded to 0.4 to 0.5 mm and for air mix the air flow should be 2.25 m^3/min/m^2 at 0.55 bar.

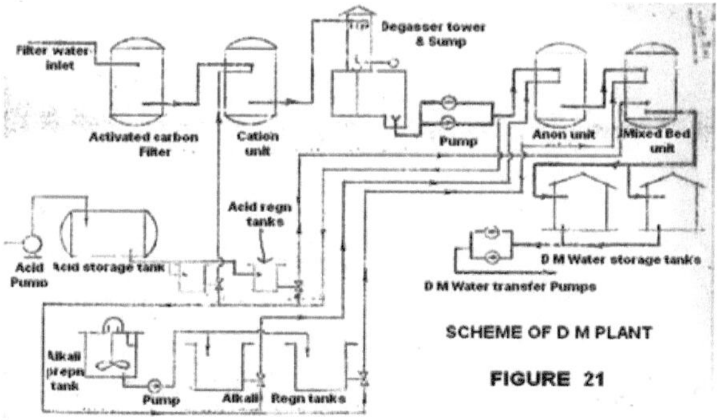

SCHEME OF D M PLANT

FIGURE 21

A conventional flow scheme of a DM plant:

A conventional DM plant (Fig. 21) consists of a series of sand and activated filters followed by cation unit, degasser, anion unit, mixed bed unit along with acid & alkali regeneration and bulk chemical handling systems. For further polishing and removal of colloidal silica in DM water, ultra filtration system is used. General arrangement drawing of an ultra filtration system is shown in the diagram below.

Degasser / Decarbonator

To make a demineralisation plant operation more cost-effective, i.e. to reduce the amount of caustic soda for regenerating anion resin and to have another weak acid, viz., silicic removed more effectively by it, it is necessary that the bulk of carbonic acids as produced from cation exchange resins are removed mechanically through a degassifying arrangement. The carbonic acid easily breaks into carbon dioxide and water:

$$H_2CO_3 \rightarrow H_2O + CO_2$$

and the gas carbon dioxide is then removed from the system by the principle of solubility of gases under Henry's law of partial pressures. For the purpose, a forced draft degassifier tower as shown in figure is usually employed in a TPS

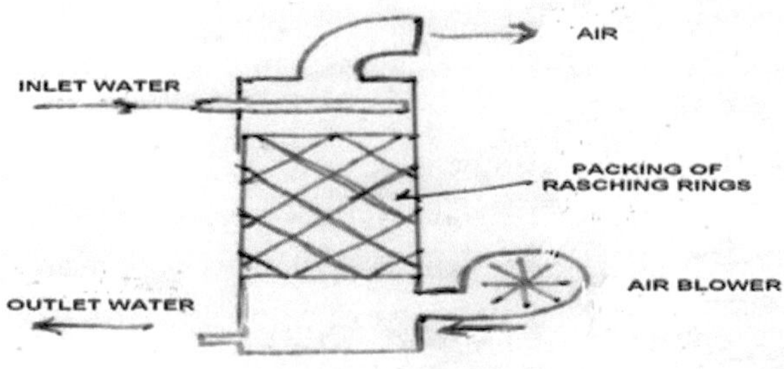

FORCED DRAFT DEGASSER TOWER

FIGURE 22

The decationised water as obtained after cation exchanger containing carbonic and other acids are passed on to a tower having a suitable packing height of trays consisting of raschig rings usually of polypropylene material, which breaks the water into small droplets and in this form are allowed to travel down through a stream of air flowing in the opposite direction. The flowing water is thus stripped from CO_2 and is collected in a storage tank usually provided underneath of the tower for its progressive use to anion exchanger resins. In view of the saturation limit of CO_2 in water being 5 ppm, the decationised water can easily be stripped to have remnant of this amount only. To reach however a lower level up to 2 ppm calls for a disproportionate amount of extra packing and very attentive operation of the tower with purified air having local disturbance of CO_2 in the air. A vacuum degasser (same type of tower but operated on suitable vacuum) can however reduce the level of CO_2 to 1 ppm along with the oxygen also removed almost entirely (<0.5 ppm). This type of degasser is rarely in use in a TPS in view of its techno economics & operational difficulty.

Degasser tower (DGT) design is relatively simple. The tower diametre is arrived at design flow rate of water maximum up to 60 $m^3/m^2/h$. Standard designs of DGT incorporates 50 mm diametre pall rings as a packing, taking into account, the factors such as height of transfer unit, number of transfer units and safety margin. The packing height is fixed based on designer's experience. Residual carbon dioxide (CO_2) at outlet of tower is generally 5 mg/l (which is soluble in water).

As a convention, however, a degassifier arranged in a DM plant is considered at flows over 100 gpm and alkalinity over 100 ppm. Thus, sequence of a DM plant in general is activated carbon filter, cation exchanger, degasser, anion exchanger and finally mixed bed, the effluent of which is led to DM water storage tank well protected from atmospheric CO_2 contamination. This sequence is however modified as per demand by the existent raw water quality or in an attempt to save the regenerant chemicals and having more efficient operation.

Other Types of DM Plant

DM plant design varies from one place to another depending on (a) source of water available (b) ionic load of water (c) treated water quality required (d) regenerant chemical availability. Accordingly different purity water can be obtained as below in terms of its conductivity, which is taken as ready test of purity of water:

Demineralised water	10 micromhos/cm
Mixed bed polished water	1 micromhos/cm
Ultra pure	0.1 micromhos/cm
Theoretically pure	0.058 micromhos/cm

Where very pure/ultra pure quality or high silica purity are required, installation of mixed bed after anion exchanger is almost certain to remove remnants of ions from the deionised water. To remove colloidal silica, installation of ultra filtration unit after the mixed bed is on card. Where dissolved salts are more than 500 ppm, reverse osmosis plant followed with a DM plant have been found economical.

To remove high dissolved salts, single bed exchangers of cation and anion are uneconomical. For this, weak acid cation exchange resin bed in series with strong acidic cation exchange bed and similarly weak basic anion exchanger bed in series with the strong one are more economical. However, where both weak and strong anion exchanger beds are provided, degasser unit can be installed after weak base exchanger. This sequence of installation enhances the efficiency of the degassification, in view of only weak acids dominating in the decationised water. The following are the different flow sheets of water demineralisation generally employed in TPS with respect to available raw water quality at site:

Flow Sheets

1. Raw water having low TDS (<100 ppm) & low alkalinity (<50 ppm) with low silica (SiO_2) / total anions (TA) in terms of% (<5).

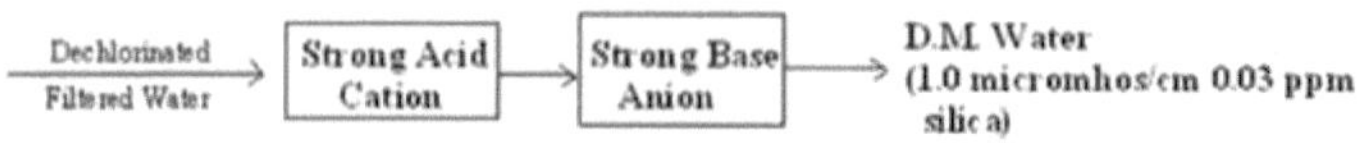

2. Raw water having quality above, but with silica ratio between 5 and 10.

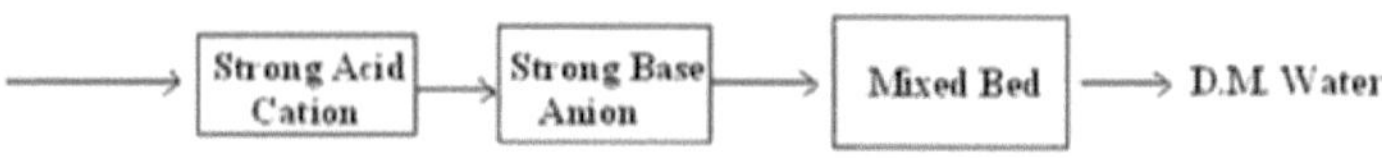

3. Raw water having TDS (<200 ppm), medium alkalinity (<150 ppm) with medium SiO_2 /TA ratio (<20%, above which counter current regeneration operation is envisaged.

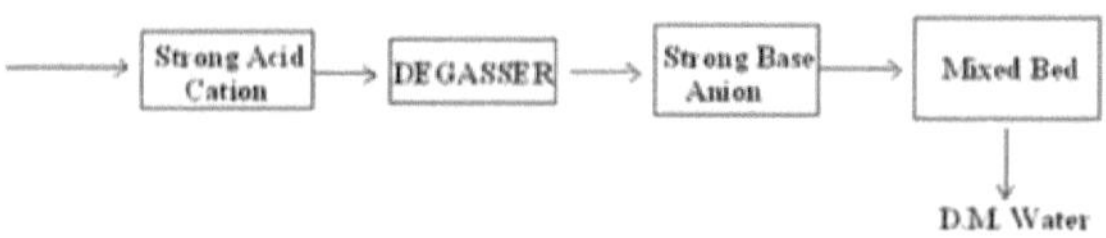

4. Raw water having quality as at (3) but having colloidal SiO_2 (>2 ppm).

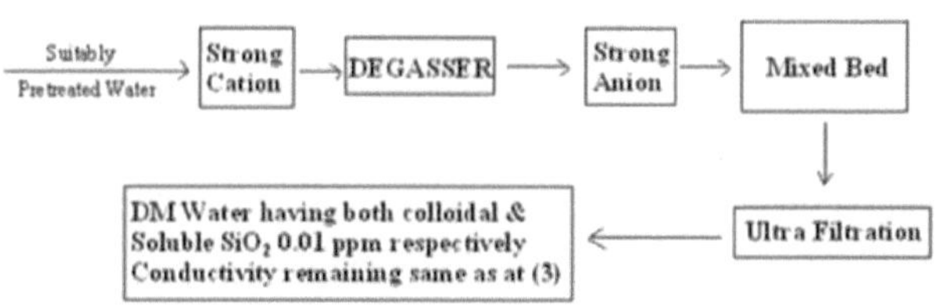

NOTE: Here the pretreatment of water needs specialised clarification/ filtration system, including use of coagulation aid, with increased detention time and slow filtration with multimedia filter.

5. Raw water having medium TDS but high alkalinity (>250 ppm).

6. Raw water having both high TDS (>250 ppm) and high alkalinity.

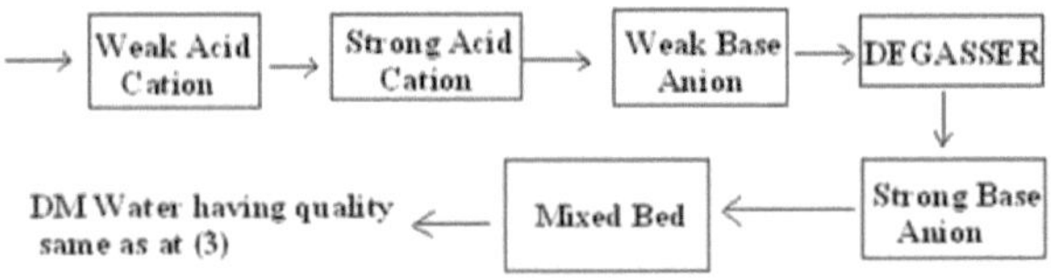

7. Raw water having very high TDS (>500 ppm) and very high alkalinity (>500 ppm) (*).

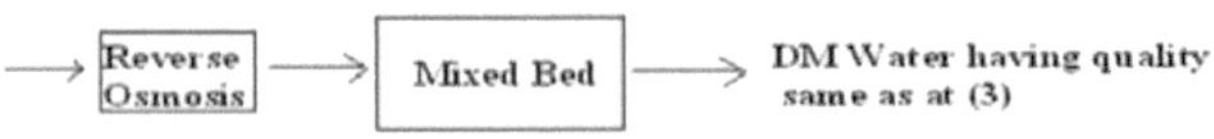

(*) Detailed pretreatment as at (4)

8. When raw water is brackish water / sea water.

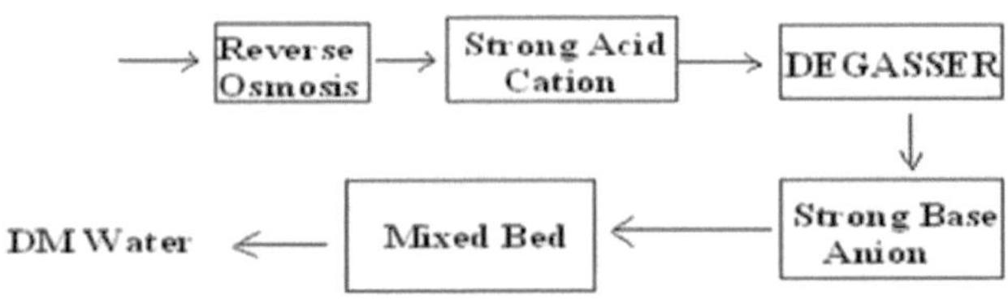

(*) Detailed pretreatment of water as at (4) and including use of suitable coagulant & biocide

9. Additional flow sheet of DM plant as practiced in some of TPS DM water quality as practiced in TPS.

Boiler operating pressure kg/cm^2	Below 50	Between 50 - 100	Between 101 - 150	Above 150
Conductivity µS/cm@25 ^{0}C	< 1.0	0.3 - 0.5	0.2 - 0.3	0.1 - 0.2
pH@ 25 ^{0}C	7.0 +/ 0.3	---	7.0+/--0.2	---
Total silica, as SiO_2 ppm	< 0.02	---	< 0.02	---

For a 500 MW TPS, DM water quality is invariably produced of 0.1 micromhos/cm conductivity@ 25 ^{0}C and pH of 7.00.2±. This further has following characteristics:

Total dissolved salts	50 ppb
Total ionised salts	25 ppb
Dissolved iron as Fe	5 ppb
Dissolved silica as SiO2	20
Dissolved copper as Cu	5 ppb
Colloidal impurities	5 ppb
Total organic carbon	60 ppb

Operation of DM Plant

The first and foremost thing to remember before operating a DM plant is that inlet water quality to the DM plant must be of a very good quality as below:

1. The water should be clean, cold, chlorine & colloids free.
2. Ratio of organic matter to dissolved salts in the water (OA/TA) must not be greater than 0.015: OA = oxygen absorbed at 37 ^{0}C; 4 hours.
3. Heavy metals content, viz., Fe, Al, Mn, Zn, etc. should not be greater than 0.1 ppm.

Further, the basic steps in operating a DM plant for each of the exhangers are as below:

1. It is followed bybackwash not only classifies the bed but also removes dirt & resin fines. The same should be done with good quality filter water with a suitable flow (usually at water velocity of 10 - 15 m/h) so as to have bed expansion of 40%. The backwash in counter current regeneration is occassional. It is always followed with double quantity of chemicals during regeneration. During, each normal regeneration, there is a provision of surface wash of the resin bed. In mixed bed, backwash not only removes dirt & resin fines, but also facilitates separation of cation and anion resin in two distinct layers to ease regeneration separately.
2. Injection of regenerants should be of specified strength, flow & duration through ejector or pump as the case may be. During regeneration, resin must be in compact state. Hence, in counter current mode of regeneration, downflow from top of vessel is given either with water or air. In MB regeneration, upflow is given while regenerating anion resin & downflow while regenerating cation resin.
3. Suitable rinsing has to be done after regeneration to remove excess regenerant & to achieve specified quality. In counter current regeneration, rinsing is in two stages – first slow rinse in the direction of flow of regenerant under same flow & then fast rinsing coccurently. In mixed bed, rinsing is in three stages:

   a. Anion resin rinse to bring down the conductivity to 50 micromhos/cm approx.
   b. Cation resin rinse to get FMA value of approx. 50 ppm.
   c. Finally fast rinsing usually at the service flow rate, after mixing the resin with air.

The following rinse characteristics determine the completion of rinse in general:

Cation exchanger: FMA & EMA should be virtually equal. Also, hardness nil & sodium slip to be as per design data.

Anion exchanger: Conductivity should be less than 10 micromhos/cm, pH should be around 8.3 and silica should invariably be not greater than 0.2 ppm.

Mixed bed exchanger: Conductivity should be less than 0.2 micromhos/cm, pH 7.0 +/-0.2, and silica less than 0.02 ppm.

Above all, degasser outlet water quality should be free from dust & dirt and CO_2 should be around 5 ppm.

After the rinse is complete and the exchanger bed has to be taken into service, it is necessary to know the ionic loading on the individual beds and the useful throughput through it as below:

Total Cation	=	Total Hardness + Sodium
	=	AIK + EMA
Anion Load	=	AIK + EMA + SiO_2 (When no degasser)
		EMA + SiO_2 + Res. CO_2 (When there is a degasser)

These ionic loading can be made available from the current analysis of water inlet to the individual beds, but the exchange capacity of the individual resins can be calculated out from the regeneration graphs of the resin (as available in manufacturer's pamphlet) and the analysis results of inlet water. Then putting these values and the resin quantity available in the exchanger bed in the following formula, the expected throughput between the two regenerations, is calculated out.

Exchange Capacity (g/l) x Resin Volume (l) = Output (m^3) x Ionic Load (ppm)

From the regeneration graphs, current water analysis results and the regeneration level taken for the resins, leakage of Na or SiO, as theoretically obtained, need to be also verified with the laboratory test results, which are generally found much lower than the theoretical results.

OPERATION: The individual bed then needs to be operated till the above useful throughput is obtained or till leakage of sodium or silica from cation or anion respectively exceed the design data, whichever event is noted earlier. The exchanger unit is then isolated for regeneration.

However, mixed bed is to be operated as per design throughput capacity (and not on regeneration graph/ionic loading, etc.) or silica/conductivity increase, whichever is noted earlier. However, mixed bed is generally designed on the basis of a fixed resin quantity (with each resin bed depth of not less than 500 cm) or on the silica exchange capacity to exchange silica as available from the preceding anion exchanger unit to a specified minimum.

DM water is then stored in a suitable DM water storage tank, well protected from atmospheric CO_2, dust and oxygen. Also all regeneration waste water needs to be neutralised properly in a pit & then discharged through main waste water treatment plant meeting the quality as per pollution control board norms.

It is now customary to use decationised water for regeneration of cation bed and DM water for anion & mixed beds. All precautions are taken to avoid fouling of resins during operation.

Overcome Operating Difficulties by Knowing the Actual Causes

Problems Faced	Possible Reasons	Check for Actual Causes for Remedy
(A) Low Capacity	(1) Distributor blockage	(a) Shift in bed support (b) Violent backwash with high pressure, excess service flow (c) Bed chanelling (d) Hills & valleys in bed (e) Entrained air
	(2) Bed deposits	(a) Silt, mud & clay, etc.(b) Resin fines, CaSO4
	(3) Bed fouling	(a) Iron & CaSO4 (cation resin) (b) SiO2, iron & organics (anion resin)

Problems Faced	Possible Reasons	Check for Actual Causes for Remedy
	(4) Bed material	(a) Decrosslinking of resin
		(b) Attrition of resin
		(c) Damage of resin due to oxygen or Cl2 confirm from data
	(5) Temperature decrease	
	(6) Over run in previous run	
	(7) Ionic load increase	
	(8) Excessive rinsing	
(B)Low Pressure Drop	(1) Reduced bed depth	Confirm
	(2) Decreased flow rate	Confirm
	(3) Increased temperature	Confirm
(C)High Press drop	(1) Bed deposits	Confirm
	(2) Excessive flow rate	Confirm
	(3) Unclassified bed	Confirm
	(4) Throttled valve opening	Confirm
	(5) Temperature decrease / viscosity increase	Confirm
(D) Resin Loss	(1) Free board too shallow	Check
	(2) Excessive backwash pressure	Check
	(3) Entrained air/release of gases during backwash	Check
	(4) StrainerBreakage	Check

Problems Faced	Possible Reasons	Check for Actual Causes for Remedy
(E) Poor Treated Water	(1) Over exhaustion	Confirm
	(2) Excess/low flow rate	Confirm
	(3) Poor inlet water quality	Confirm
	(4) Valve leakage	Confirm
	(5) Poor regeneration of the other reasons under this column other reasons under this column	(a) Impure grade chemicals (b) Less quantity of regenerant used (c) Less strength of regenerant taken Investigate
	(6) Any poor (MB only) corrosion products	
	(7) Resin separation	(a) Anion resin density increased (b) Excess or low regeneration flows
	(8) Improper mixing of resins (MB only)	(a) Cation resin density decreased due to oil adsorption & attrition of resins (b) Clumping or turbulence in central drain line

Problems Faced	Possible Reasons	Check for Actual Causes for Remedy
(F) Colour Throw	(1) Resin deteriotation	(a) Resin storage at high temperature
		(b) Freezing & thawing of resin
		(c) Inlet water contamination continuously with high iron, bacteria, algae, etc.

Operating Points to Remember in DM plant Operation

1. Exchange capacities as given under the graph, by the resin manufacturer, are obtained under following ideal conditions, and the given values are not guaranteed, but are for guidelines:
   a) Pilot exchanger vessels have small diametre/high bed depth ratio (300 mm diametre & 1400 mm bed depth: called as an ideal ratio)
   b) Absolute perfect distribution through the resin bed
   c) Maximum contact time for the regenerant applied
   d) Favourable kinetic conditions by using high TDS water
   e) No organically fouled anion resin. Silica slip through it is based on zero sodium slip from cation resin. It is therefore desirable to downrate the exchange capacities by an engineering factor which would depend on:
      a) Raw water analysis
      b) Treated water quality
      c) Flow distribution through headers & laterals
      d) Kinetics favourability
      e) Ageing of resin & quantum of deterioration
      f) Temperature effect including sunlight on one side of unit
      g) Chlorine content
      h) Regeneration techniques.

1. Cation & anion exchanger resins are usually downrated by 10 - 15% when used in cation & anion beds, but the same when used in mixed beds are downrated up to 40% due to poor separation of resins and also necessary to regenerate the two resins simultaneously even when only one is found exhausted
2. Although cation & anion resins have quite similar densities 1.27 & 1.07 respectively, their effective density in water on archemedes principle is 0.27 & 0.07 respectively, which is very substantial difference, quite sufficient to get a clean separation. By using a water of density greater than 1.0, the difference in effective densities can be enhanced & hence in difficult situations sometimes 12% NaOH solution is used for separation of the two resins in the mixed bed.
3. Silica leakage below 0.02 ppm can be achieved with mixed bed, only when the bed is not overloaded (greater than 50% of the exchange groups) with silica and if the cycle time is kept below one week. This is because of peculiar characteristics of silica as regards its solubility & polymerisation ability. If anion resins are loaded with 2 g/l SiO_2 during exhaustion, precipitation of silica may result while regenerating with NaOH. Hence, anion resin must be given upflow during regeneration to save cation resin from spoiling. Also, SiO_2 leakage from anion unit is guaranteed (as per graph) assuming zero Na slip through cation unit. For every 1 ppm of Na slip (as $CaCO_3$), the residual SiO_2 must be increased by 15%. With loss of strong capacity to 50% means more silica leakage.
4. Strong Base resins are regenerated at a regeneration ratio of 3.0 and above, in order to obtain a reasonable capacity & to keep the SiO leakage down, but a ratio of 1.3 to 1.6 is normal in case of weak base exchanger. In case of cation resins, nCa++ are held so strongly on the resins that a high regeneration ratio (2.5 - 3) are needed to knock it off, but even with Na+, which is less strongly held, a ratio less than 1.5 is impractical. [Regeneration Ratio = gm equivalent of regenerant input / gm equivalent remained on resin]. If the anion bed is to be left in standby condition, it is better to leave in the regenerated state. This will avoid polymerisation of silica in the resin bed. Strong acid cation bed, however, if left on regenerated state for a longer time may degrade its sulphonic group due to air/oxygen oxidation.

5. Leakage of sodium through cation bed will cause leakage of NaCl through weak base exchanger i.e. 1 ppm Na leakage through cation bed will increase conductivity of weak anion effluent by 2.2 micromhos/cm. Similarly, 1 ppm Na leakage through cation will increase strong anion effluent conductivity by 5 micromhos/cm and SiO_2, by 15%. Also, since leakage here is converted to NaOH, more 'Na' slip from cation may regenerate anion resin and give slip of released ions from the bed (viz., Cl-, etc.).

6. The purity of water treated through anion bed or mixed bed is checked on basis of its conductivity and silica content. The pH is another parametre to be monitored, but it does not indicate exact purity of water. The pH measurement of high purity water is difficult due to absence of adequate amount of electrolytes. Special precautions have to be taken for measurement of pH of mixed bed treated water.

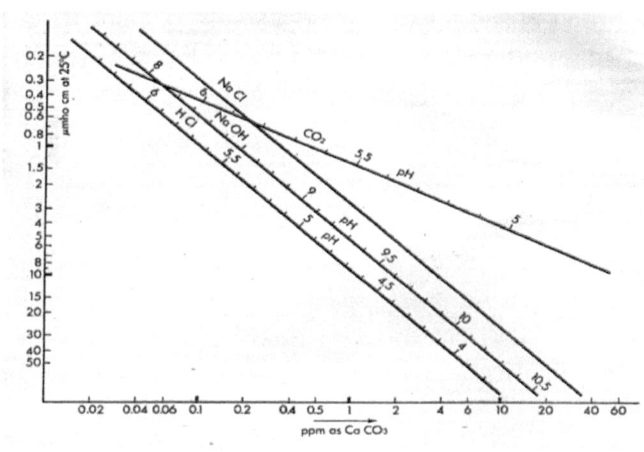

Purity of water is 0.1 micromhos/cm conductivity will have:

Total solids	50 ppb
Ionised constituents	25 ppb
Fe	5 ppb
Cu	5 ppb
SiO_2	10 ppb
Colloidal impurities	5 ppb

This quality can be produced, provided the MB influent quality has less than 10 micromhos/cm conductivity and less than 10 ppb crud loading. It is therefore better to reclaim the excess MB rinse.

7. Counter current regeneration gives low slip of sodium during the cycle, but length of ion exchange zones is short at the end of the run, leading to immediate high increase of slip. In counter current regeneration of anion bed, there is possibility of spent NaOH containing SiO_2 to pass through neutral or acidic zones giving polymerisation/precipitation of SiO_2. Hence, either 2% NaOH is used or in gradient from 2 to 4% for the regeneration. Regenerant flow is generally kept to 2 BV/h. (This case is pertaining to weak base & strong base anions thoroughfare regeneration).

8. Swelling of strong acid cation resins from Na^+ to H^+ form is approximately 7% and that of strong base anion resins from Cl- to OH- is about 12%. All types of strong base resins tend to loose part of the strong base group over a period of time. The result is that anion resins tend to loose its capacity to absorb weak acids like CO_2 and also silica. The reeplacement of resins depends upon the irreversible loss in its operating capacity.

9. A collecting system in ion exchangers is to be designed properly to ensure uniform collection of treated water during service step and regenerants during regeneration step.

10. Silica in the form of H_2SiO_3 is exchanged by anion resin, which is monovalent and hence its equivalent weight is taken as 60 for all calculation purposes. As silica concentration within a resin bed increases and as the pH within the bead's environment decreases, silica within the bead will polymerise, viz., x. SiO_2 -> (SiO_2) , then regeneration step with NaOH becomes more of a dissolution process than merely an ion exchange, viz., adsorption and generally it is made feasible by hot regeneration (50 ^{0}C). This polymer formation within the bead increases on ageing. Hence to avoid polymerisation, the standby anion resin column should be kept in regenerated state or loaded with SiO / TA <50%. On polymerisation, silica is converted to dimers, trimers, tetramers, etc.

11. Water is from cation exchanger is always acidic due to presence of excess of H^+ ions and similarly the water leaving an anion exchanger is always alkaline as long as the unit is not exhausted. For this reason, it is

impossible to treat neutral water by exchanging anions first, because the high pH, which the anion resin would generate, would precipitate the Ca^{++} & Mg^{++} in the water. Reverse deionisation (viz., first anion and then cation) can however be feasible with very low hardness water and this type of system can also be employed when anion resins are to be used as organic trap, desilicisers (usually on high sodium water).

12. Filming amine and coagulant aids foul cation exchange resin and detergents foul anion exchange resin. Polyphosphate also may inhibit the ion exchange reactions in DM plant. Ions as a matter should diffuse freely into the resin matrix. Obstruction like oil, grease, organic fluids & solid compounds insoluble in water make the resin surface water repellant. Weak base anion resins do not foul with organics but they develope excessive swelling between regeneration & exhaust forms.
13. Both cation & anion resins are resistant to NaCl, HCl, H_2SO_4, NaOH, but strongly basic anion resins react with NaOH depending on its concentration and is accelerated by increase in temperature & presence of oxygen. This reaction leads to separation of amines, which can be identified by means of odour evolved during regeneration and weakly basic groups are formed separating nitrogen group. The service life of anion resins can therefore be increased by regenerating with dilute & cold lye. The lye should however be free from bicarbonate & carbonate, which would otherwise reduce the anion capacity. This should not also contain more than 0.001% of oxidising substances like sodium chlorate, etc.
14. Low crosslinked resins regenerate more efficiently than higher ones, but an optimum crosslinking is a must for physical stability of resins. Iron acts as a catalyst in oxidation and breakdown of normal sulphonate styrene divinyl benzene copolymer, which will cause fouling of cation resin. However, high crossed linked cation resins are more resistant to oxidation. For mixed beds, resins should be selected from 10 - 12% DVB cross linked, if gel type resins are chosen or prefer macroporous resins.
15. Wastewater generated with regeneration is generally 11 BV and with HCl/ NaOH is 10 BV. The break up with H_2SO_4 could be as below:

Backwash	2 BV
Injection	2.8 BV
Rinsing	6 BV

Plant Monitoring

Pretreatment

Locations	Parametres	Unit	Frequency
Inlet to aerator	pH	-	Once a day
	TSS	mg/l	
	Turbidity		
	Organic matter	mg/l	
	Hardness	mg/l	
	Alkalinity	mg/l	
	Chlorides	mg/l	
	Sulphates	mg/l	
	Nitrates	mg/l	
	Silica	mg/l	
	Heavy metals	mg/l	
Clarifier outlet	pH		Once a shift
	Free residual chlorine	mg/l	
	TSS	mg/l	
	Turbidity	NTU	
Filter outlet	TSS	mg/l	
	Turbidity	NTU	
Carbon Filter outlet	FRC	mg/l	

Demineralisation

Locations	Parametres	Unit	Frequency
SBA outlet	pH		Once a shift
	Conductivity	µS/cm	
	Silica	mg/l	
MB outlet	pH	-	Once a shift
	Conductivity	µS/cm	
	Silica	mg/l	

Other Processes of Demineralisation Membrane Processes

When DM plant operation becomes cost-effective with high TDS in raw water, production of pure water is also achieved by membrane processes - the water to be treated by such technology involves fresh water, brackish water, sea water, waste water and other water in different industries.

Membrane technology is advancing quickly, becoming gradually less expensive with improving performance & life expectancy. Rather, demineralising water having more than 200 ppm TDS now, reverse osmosis plant under membrane technology prior to mixed bed resins (DM plant) has proved to be more suitable techno economically.

Common membrane processes include RO, electro dialysis reversal (EDR) & ultra filtration (UF). The UF however is employed for efficient removal of non-ionised organics, colloidal particles & other particulates in place of removing ions as done by RO & EDR.

In all these processes, it is however essential for vigorous pretreatment of raw water as fouling control measures of the membrane which include micro filtration of water & antiscalant dosing apart from suitable clarification ahead of this. This is required in view of membrane processes producing a concentration gradient of dissolved salts approaching membrane surfaces.

Reverse Osmosis (RO) Plant

Reverse osmosis, the basic principle of which is illustrated in, derives from the fact that the direct or natural osmosis is a reversible process. Thus, to a saline solution, if a hydrostatic pressure in excess of the osmotic pressure is applied, it is found that pure water flows in the direction opposite to that of osmosis phenomenon, and that the salts are retained by the membrane, which is of semi-permeable in nature, allowing water to flow only in one direction. While the membrane allows pure water to pass through, 95 - 99% of inorganic salts present in it in the form of solution are retained (viz., salt passage, abbreviated to 'SP' lies in the 10 to 1% range). Also, 95 to 99% of organic constituents and almost cent percent of most colloidal particles (bacteria, viruses, colloidal silica, etc.) are retained.

Fig.24 Basic Principles

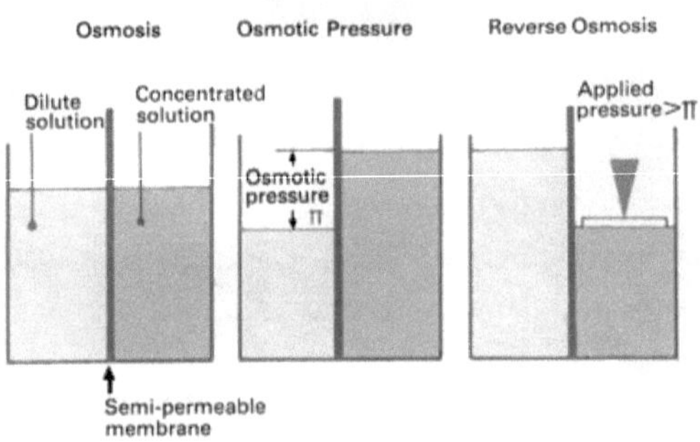

Membranes are assembled in units called modules, and the simplest practical arrangement is to connect together in series ahead of which a high pressure pump is installed. The feed water passes through the membrane under the action of the pressure collected as apermeate, while the concentrated reject is extracted continuously through a flow control valve. The ratio between the flow rate of the purified water & that of feed water is called conversion factor or recovery and is expressed as percentage. Finally to obtain a satisfactory flow of purified water it is advisable to apply a pressure well in excess of osmotic pressure to take well care of polarisation caused by concentration of the reject. A typical flow diagram of RO plant is shown below:

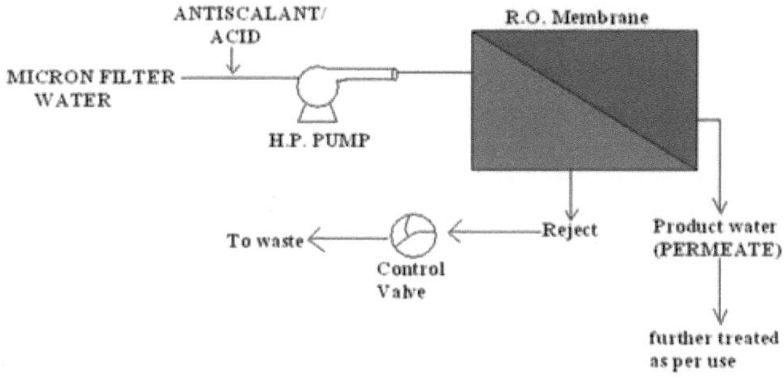

Detailed scheme is as below:

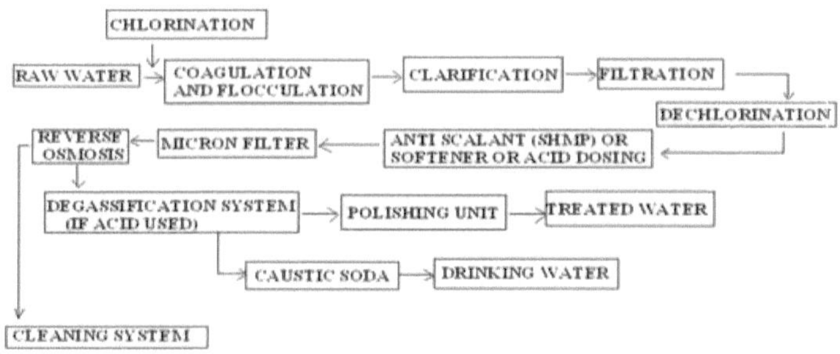

There are today two types of membrane in the market. a) Cellulose acetate (A mixture of mono, di and triacetate). These give a high flow rate per unit surface area and are used in the form of tubes, as spirally wound flat sheets and more recently as hollow fibre. These membranes are subject to hydrolysis & needs strict control of pH & temperature.

Polyamide membrane, on the other hand has a lower specific rate of flow. They are in the form of hollow fibres so as to achieve the maximum surface area per unit volume, 15 times that of spirally wound membranes. These membranes posses outstanding resistance to chemical & biological agents.

Nowadays thin film composite membrane are in practice, which consists of a thin layer of one polymer cast upon a thicker supporting layer of a different polymer (either cellulose acetate or polyamide aramide polymer). These are generally used in higher rejection permeates, but these membranes are more difficult to manufacture consistently.

Membrane modules may be staged in various design configuration producing the highest quality permeate with the least amount of waste water. The staging is such that reject from one stage is passed to second stage for getting pure water (permeate). Modules are put in a cartridge of normally 4 or 8 inches diametre & about 3 feet long.

RO membranes do not remove dissolved gases such as Cl_2, CO_2, O_2, etc. due to their molecular porosity. Polyvalent ions will be held back first than monovalent.

Fe & Al should not reach the membrane, which foul the membrane surfaces quickly. To avoid calcium to exceed their solubility & precipitate at membrane surfaces, acid dosing is required to control pH at inlet to membrane. Similarly other sparingly soluble salts like silica, phosphate, zinc, hexavalent chromium, oil and any particulate matter also need specific attention to be removed at clarification/ filtration/micron filter stages. Micron filter at inlet to membrane is recommended of 5 micron size.

General engineering norms of determination of pumping pressure is 1.5 times of osmotic pressure, but actual operating pressure could be as high as 5 to 10 times of osmotic pressure. Brackish water RO system may operate at 8 to 30 bars, but sea water system at 35 to 70 bars. Nowadays, low pressure membranes have also been manufactured which can take care of raw water having TDS of as low as 190 ppm. This viability retrofits a RO step upstream of demineraliser in case of handling moderately high TDS raw water. A RO plant can be designed to operate at recovery of 90% or more, but many systems installed so far still operate at only 75% recovery.

As per literature, membrane module flow is defined at 25 ^{0}C. This increases by about 3% for each degree above 25 ^{0}C & decrease by 2.8 to 3% for each degree below 25 ^{0}C. The rate of flow through membrane is directly proportional to the effective pressure (differential pressure between applied pressure & osmotic pressure). The osmotic pressure of different salts as found in water can be found as below:

Salt	mg/l	Os. Press (Bar)	Salt	mg/l	Os. press (Bar)
NaCl	35,000	27.86	NaCl	1000	0.79
Na_2SO_4	1000	0.42	$MgSO_4$	1000	0.25
$CaCl_2$	1000	0.58	$NaHCO_3$	1000	0.89
$MgCl_2$	1000	0.67			

Loss in performance of a RO plant means mainly: (1) Loss of flow - This is attributable to mainly fouling (2) Loss of rejects (giving higher permeate conduct, which should not increase greater than 15%) - This is either due to fouling or degradation of membrane or an 'O' ring leak.

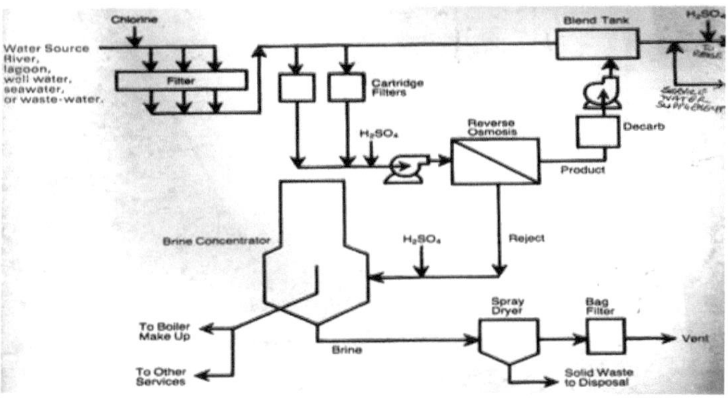

Waste Water Management — Zero Discharge

To approach zero discharge from RO plant, the reject is led to a vertical tube falling film evaporator with mechanical recompression of vapours (as shown above), dosed with scale control agents.

The overall water recovery of 98 to 99% can be achievable by combined operation of RO & VCE Plants. Distillate quality is of 5 - 10 ppm TDS.

As a precaution to membrane, it has to be kept wet always after initial wetting and first hour of permeate needs to be rejected. When operation of membrane is not required, the same is needed to be preserved suitably as per guidelines of manufacturer. Generally, the preservation is done with 1.5% sodium meta bisulphate (food grade).

Electrodialysis Reversal

Electrodialysis (ED) processes transfer ions of dissolved salts across membrane leaving purified water behind. A negative electrode (cathode) attracts cations and a positive electrode (anode) attracts anions & thus systems are compartmentalised in stacks by altering cation & anion transfer membranes. The principle of *ED* is as illustrated in following figure:

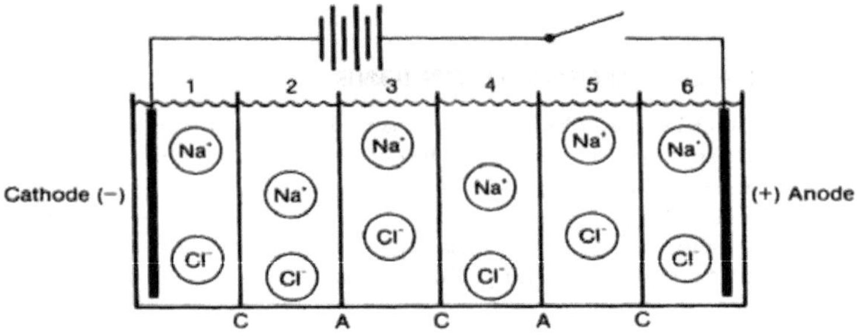

Based on the principle, ED system is made unidirectional ie. ionic movement is in one direction only (cations moving towards a fixed cathode & anions moving towards a fixed anode), but in such a system chemical dosing is usually required to prevent formation of scale caused by precipitation of CaCO & CaSO on membrane surfaces. Polarisation effect on the electrodes are also to be looked into for cleaning.

Electrodialysis

A = Anion transfer membrane C = Cation transfer membrane

Both are electrically conductive and essentially water impermeable under pressure.

Selectivity of membranes is generally above 90% for a solution less than 0.5 - 0.7 N, but decreases at higher ion concentration. Membranes for brackish water generally work at 88% current efficiency.

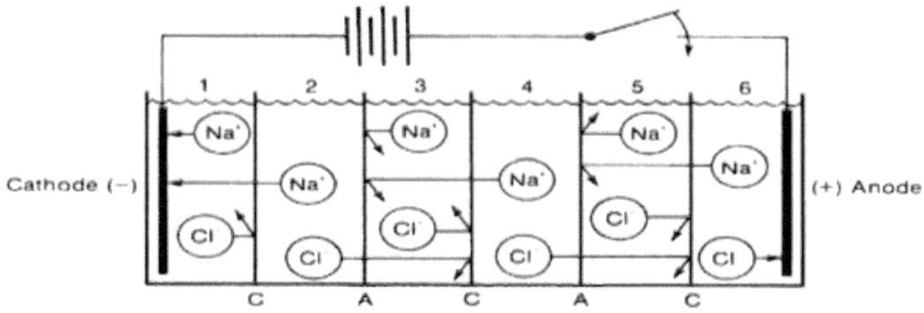

On applying DC potential, compartments 2 & 4 have been depleted of ions and the ions are concentrated in compartments 3 & 5. Chlorine gas, oxygen gas & H+ ions are produced at anode (compartment 6). Hydrogen gas & OH- ions are produced at Cathode (compartment 1)

Electrodialysis reversal (EDR) processes operate on the same principles as ED, but here operation reverses system polarity (typically 3 to 4 times per hour). This reverses the direction of ion movement within membrane stack, thus controlling film and scale formation. This reversal, therefore, stops the build up of concentrated solutions on the membrane & thereby reduces the accumulation of inorganic & organic deposition on the membrane surfaces.

However, effect of this reversal is that the concentrate steam remaining in the stack whose salinity is higher than the feed water must now be desalted. This creates a brief period of time in which the product water salinity is higher than the specified level. This slug of water is known as 'OFF SPEC PRODUCT'.

Because of reversal, no flow compartment in the stack is exposed to high solution concentration for more than 15 to 20 minutes at a time.

ED & EDR however require high powered direct current electrical fields to induce ion movement. These processes do not remove particulate contaminants or weakly ionised contaminants like silica thus, requiring enormous power consumption at only low flow rates obtained due to limitation of cell size. Also, spacers generally do not work above 47 ^{0}C.

A typical EDR flow diagram is shown in the figure below. EDR like electro dialysis works with a water having 400 to 3000 ppm of TDS.

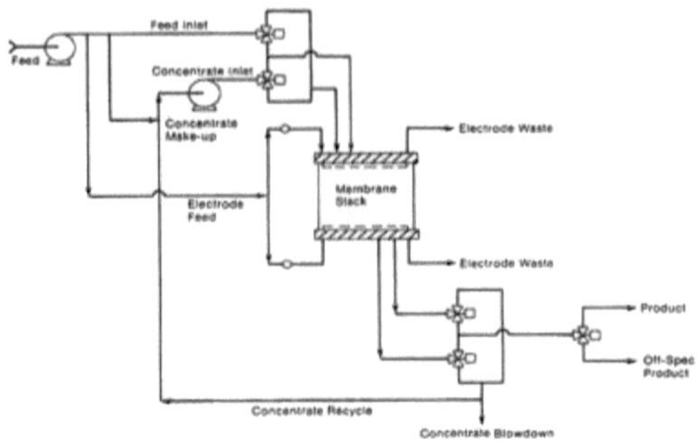

Electro-Deionisation (EDI)

This is another method of ion removal from water. There are two operating regimes for EDI devices: (a) Enhanced transfer and electro regeneration - In this, the resins within the device remain in salt form. In low conductivity solutions, the ion exchange resin is orders of magnitude more conductive than the solution and acts as a medium for transport of ions across the compartments to the surface of the ion exchange membranes. This mode of ion removal is only applicable in devices that allow simultaneous removal of both anions and cations, in order to maintain electroneutrality. (b) The second operating regime for EDI devices is known as the electro regeneration regime. This regime is characterised by the continuous regeneration of resins by hydrogen and hydroxide ions from the electrically induced dissociation of water. This dissociation preferentially occurs at bipolar interfaces in the ion-depleting compartment where localised conditions of low solute concentrations are most likely to occur. The two primary types of bipolar interfaces in EDI devices are resin/resin and resin/membrane. The optimum location for water splitting depends on the configuration of the resin filter. For mixed bed devices, water splitting at both types of interface can result in effective resin regeneration, while in layered bed devices water is dissociated primarily at the resin/membrane interface.

Regenerating the resins to their H^+ and OH^- forms allows EDI devices to remove weakly ionised compounds such as carbonic and silicic acids, and to remove weakly ionised organic compound. This mode of ion removal occurs in all EDI devices that produce ultra pure water. In a large scale EDI device, many of these cells are assembled together and fed in parellel as shown below:

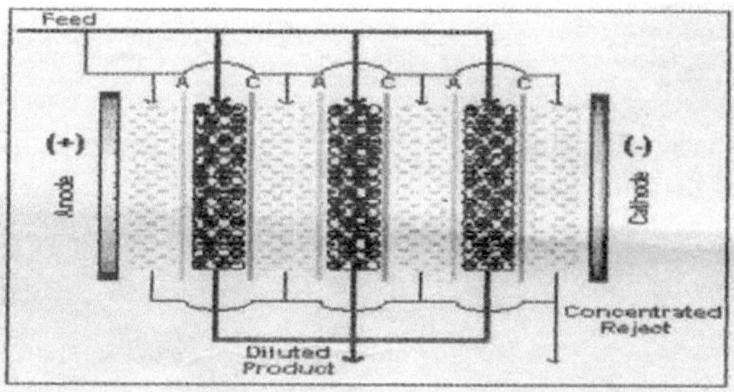

Figure 2

However, proven performance of this EDI technique in Indian TPS is yet to come.

Ultrafiltration

Ultrafiltration plant is generally required in a TPS to remove colloidal silica, if any present as 0.02 ppm in the boiler water. In view of osmotic pressure to be overcome, in this case is negligible, and the process needs to separate colloids including organic matter, colloidal SiO_2, oil, etc. & macromolecules whose molar mass exceeds the cutting limit of the membrane, a low feed pressure of water of about 2-4 bar are used for ultafiltration. However, pore size of such membrane is made much below 0.1 micron, which causes pressure differential increase in a way to produce acceptable flow rates in high range. Pressure differential in excess of 1.4 kg/cm^2 is usual. Such membrane are therefore of very thin skin (<1 micron thick) on the surface of a more porous thicker substrate (25 micron thick). This type of membrane is said to be anisotropic. A number of design techniques have been developed to take care of due filtration at high flow rate and avoiding sweeping of ultra filtration.

Membranes in view of low conversion factor used (0.5 to 5%) in the case of all heavily contaminated liquids. The macromolecules & colloids deposited on the surface show very little tendency to diffuse back into the liquid and thus specific flow rate can only be relied on which require thorough pretreatment including filtration through pores of 10 to 20 micro metre range, to avoid premature clogging of the membrane. Concentration polarisation of membrane surfaces also may affect flux rates, though less in comparison to

RO membrane. However, the specific flow rate in case of pure water (viz., to remove colloidal silica from DM water) can be achieved as 2 m^3/d per m^2 of membrane surface area at 3 bar operating pressure.extent of sweeping of the membrane wall. Least prone to clogging are of course, the tubular membranes, followed by the flat design and spirally wound & hollow fibre membranes;

The function of an ultra filtration plant & its principle can be understood by the following figure.

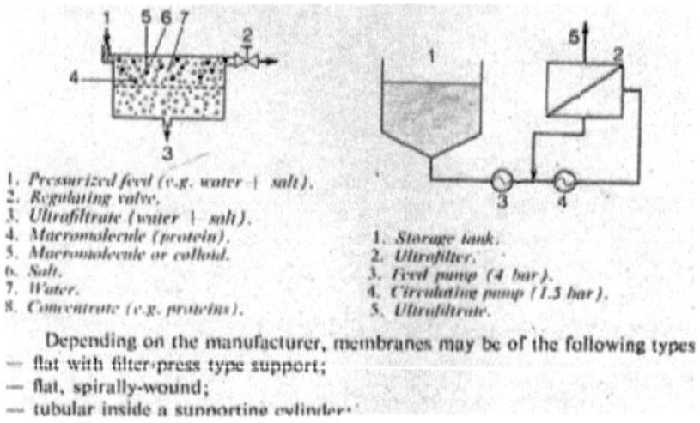

FIGURE (a) Simple ultra filtration on pure water

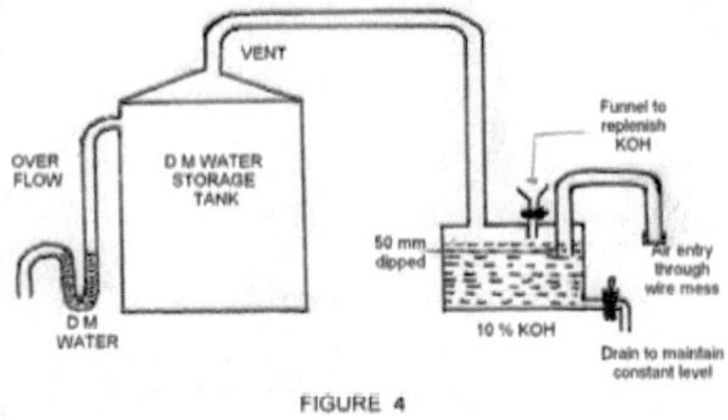

FIGURE (b) Ultrafiltration with recycle arrangement on heavily contaminated liquid (calls for high level energy consumption)

Protection of Make-up Water Tank from Atmospheric CO_2 Contamination : It is necessary that quality of DM water/condensate storage tank water is monitored with respect to CO_2, which in turn, will increase corrosion rate in hot well of condenser, feed water pipes, etc. DM water as produced from DM plant of ultra pure quality, is highly susceptible to atmospheric CO_2 contamination and thus this water as stored in DM water storage tank loses its original quality as produced and becomes sometimes sufficiently acidic unacceptable to the boiler feed. Hence, for these tanks, installation of CO_2 absorber has been found essential. Though in some utilities, floating balls are used to cover the open surface of DM water in the tank, to avoid direct contact with the atmosphere, the usual practice is to have atmospheric air (devoid of CO_2 by suitable absorption method, usually by KOH solution) in contact with water, as illustrated in figure below. It is however necessary that the vent outlet line diametre should match with the air entry line diametre to the KOH solution tank (approx. equal) and similarly the overflow line diametre with the diametre of feed line to the tank. The satisfactory working of this installation however needs to be monitored with intermittent replenishment with fresh KOH solution.

Chapter - B

BOILER FEED STEAM CYCLE CHEMISTRY

(Describes healthy operational practices of the whole boiler feed-steam cycle equipments, along with types of various causes of failure experienced including from corrosion in detail. Also deals with behaviour of different water conditioning chemicals for boiler feed water-steam cycle)

Boiler Feed – Steam Cycle Chemistry

At the outset, this chemistry, though very complex, can be started to be understood with the thermal power generation programme from an early age to the present one, as below:

Thermal Power Generation

Electric power is one of the main inputs in any industrial plant for sustained growth of a country for which various options are available in setting up power plants. The most conventional power generation in the world is through steam generators & turbines and the installation is invariably referred to as *Thermal Power Station (TPS)*. The complex nature of chemistry of water & steam at high pressures & temperatures and the very high purity requirements of steam for rolling the turbine has been actually challenging task on the part of water/ steam chemistry management, which have been continuously upgraded in view of the problems faced with respect to corrosion, deposits, etc. leading to failure of the equipments to the loss of power generation at large.

In the very early stages of thermal power development, just ordinary drinking water was used for boiler make-up. As the size of the units increased, the steam flow, pressure & temperature conditions increased tremendously for economic considerations & higher efficiency reasons. Accordingly the quality of boiler water was improved to ensure satisfactory operation of boilers & turbines, with production of ultra pure water for feed to boiler. But even this ultra pure water (virtually theoretically pure water) needs now attention for due conditioning, for trouble free operation of boiler & turbine, and envisaging for their long life. The typical unit size of the country & abroad is now 500 MW, of which water-steam flow diagram along with sampling system, for quality monitoring is illustrated in the following figures. There are other designs of boilers also including once through mostly in higher sizes and waste heat recovery boilers of different sizes, etc.

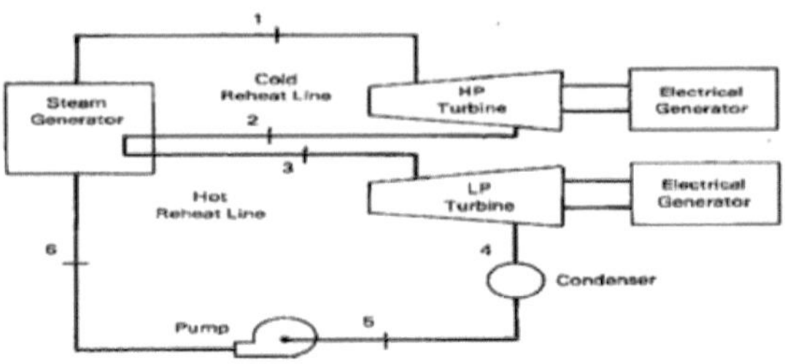

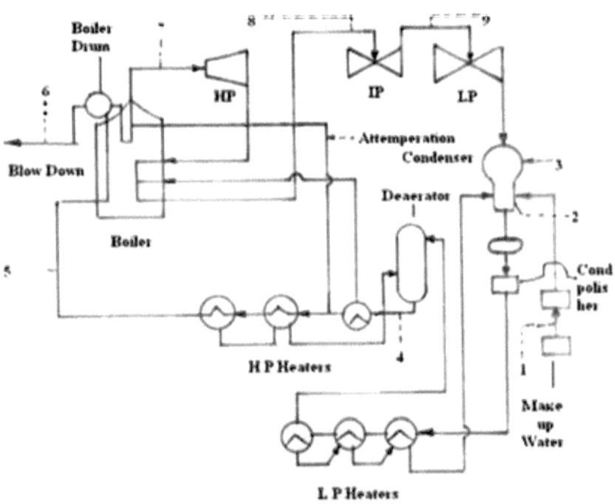

1. MAKE UP TREATMENT SYSTEM
2. CONDENSER LEAK DETECTION TRAP or HOT WELL ZONES
3. AIR REMOVAL SYSTEM EXHAUST
4. DEAERATOR OUTLET
5. ECONOMISER INLET
6. BOILER BLOW DOWN or DOWN COMER
7. SATURATED STEAM
8. REHEAT STEAM
9. LOW PRESSURE STEAM (OPERATING)

FEED WATER--STEAM SAMPLING SYSTEM

FIGURE 1

In a TPS, with the installation of condenser in line, almost all amount of steam passing through turbine is recovered in the form of water to generate steam again in boiler, but yet small quantity of water as 'boiler make-up' is required, which is though ultrapure, yet have to undergo treatment & conditioning. This make-up water in the form of DM water is usually added to the condenser and thus the return condensate & make-up water together is called as 'boiler feed water', which needs elaborate condioning before sending it again to the steam generating equipment, viz., 'boiler'. In a large size plant (500 MW & above), this condensate as mixed with make-up water is even further polished to purify for traces of both suspended & dissolved contaminants, and the plant doing so is called 'condensate polishing plant'.

Huge amount of water is however required for steam to condense in the condenser, which usually are in the form of a number of tubes, inside which suitably treated raw water (to be discussed in later chapter, i.e. cooling water chemistry') continuously flow. The expanded steam from turbine falling on the outer surface of the tubes is thus condensed to collect in the so called 'hot well', from where return condensate is formed.

In view of the most TPS employing generally high pressure boilers to generate the steam, supply of make-up water quality to these, invariably fall in the ultapure range. It is therefore necessary to be aware of, how this is further required for onward conditioning. This is revealed by the following basic qualities of pure water, which is very much aggressive at high temperature and pressure:

In case of solubility of ions makes pure water of conductivity 0.06 micromhos/cm & pH 7.0 to exhibit conductivity of 800 micromhos/cm and pH of 5.0 at boiler operating temperature and pressure (viz., 350 ^{0}C; 100 kg/cm^2). Hence, pH correction of the water is necessary to avoid acidic conditions in boiler.

It affects the dissolution of electrolytes including those of weak ones like silica, copper, iron, etc. and strong ones like alkali metals. Hence, it is necesssary to limit concentration of these to avoid deposits.

Water obeys phase distribution law and thus all dissolved salts are liable to be carried over to steam. Hence, boiler dissolved salts are to be controlled to keep steam pure. Water is very corrosive in presence of CO_2 & O_2 even in minute quantity. Also little ammonia in presence of oxygen is very corrosive to copper alloy. Hence, presence of carbon dioxide, oxygen needs to be negligible along with ammonia content. Thus, elevation of pH of the system has to be in controlled amount.

Polarisability of water molecule in contact with iron surface leads to weakening of O-H bond & gives rise to corrosion reaction by combined action of solubilisation & hydrolysis. Hence, it is necessary to have passive oxide layer on the iron surface to limit the corrosion.

Hence, there is practice of conditioning the ultra pure make-up water/ boiler feed water with suitable chemicals along with removal of most part of O_2 & CO_2 mechanically, viz., adding of make-up water by suitable spray arrangement to the condenser to make the same in droplets form, to ease the dissolved gases present in it removed through continuously running ejectors or vacuum pump as installed for the condenser to maintain proper vacuum. There is further installation of suitable 'deaerator' to deaerate the make-up mixed return condensate to remove dissolved oxygen in it, before it becomes actual feed to the boiler.

Thermal Deaeration

The following chemicals are generally used to condition the boiler feed & water itself:

Hydrazine:

It is used as an oxygen scavenger to remove remnant of dissolved oxygen from the boiler feed water by

$$N_2H_4 + O_2 \rightarrow N_2 + H_2O$$

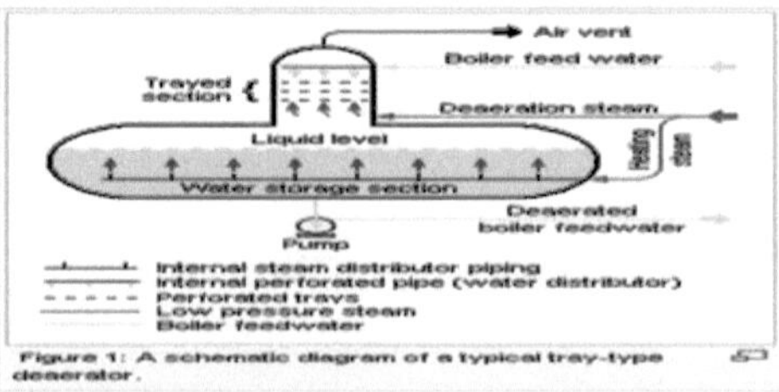

Figure 1: A schematic diagram of a typical tray-type deaerator.

In view of the very low level of remnant oxygen the reaction rate is very slow unless some catalytic effect due to presence of some metals/metal oxides are associated. Hence, it is necessary that a residual hydrazine content is always left behind in the boiler feed water to take care of remnant oxygen properly in the system. Hydrazine is generally added 2 to 4 times of dissolved oxygen present in the feed water, but not to have more than 0.02 ppm of residual in it. The more residual is likely to give unnecessary contamination of steam with hydrogen in view of hydrazine liable to be decomposed at steam temperature, if carried over to steam by phase distribution.

Hydrazine is also used as passivating agent for the mild steel surfaces at high temperature and also preserving agent at ambient temperature to save the boiler surfaces from corrosion on 'offload'.

But, since hydrazine has also been recognised as carcinogenic agent, the use of this is being limited and instead condenser & deaerator designs are upgraded to do away with dissolved oxygen. Exploring of suitable alternative for this chemical is on way with different proposal to do away with remnant of dissolved oxygen, but still no suitable alternative to hydrazine has yet been in site, for preservation of boilers, particularly when idle.

Hydrazine is usually added at inlet point to economiser during normal operation, but during start up conditions when more make-up water is anticipated, the same is dosed at the condensate pump discharge point to take care of excess dissolved oxygen suitably.

Ammonia:

Ammonia is similarly added to feed water and condensate pump discharge points to take care of suitable alkalinity in whole of the 'feed water-steam cycle' to avoid corrosion of mild steel surfaces at different temperatures & also to avoid erosion at the end point of turbine where first droplet of water appears due to condensation. Ammonia is also helpful in neutralising remnant of CO_2 in the feed-steam cycle, although the main being removed by properly storing the DM water in a storage tank facilitated with CO_2 absorber and then part by condenser & deaerator. It also neutralises other acidic constituents also, thus reducing harmful effects of acids. It also helps in keeping the silica content in the dissolved state. In view of corrosion being a function of pH, ammonia also helps in the stability function of passive protective layer of mild steel including magnetite.

Use of ammonia, however, has to be limited to get the specified pH of feed water and steam. Excess of ammonia will not only leach out copper from copper alloy from the system, but will increase the distribution of chloride to steam by formation of ammonium chloride in the system.

In some of the power stations, ammonia and hydrazine mixed in adequate proportion are dosed from the same tank at required points.

Trisodium Phosphate:

Ammonia is competent enough to maintain pH of boiler feed-steam cycle, but in case of drum type boilers where both water and steam phases exist together, ammonia escapes more into steam from the water (in the ratio of approx. 1:10). Boiler (drum) water thus remains sufficiently less alkaline than the steam. To achieve optimum pH of drum water, more ammonia dosing is likely to result more ammonia content in condensate & feed water causing concern not only for copper alloy metallurgy but also for unrequired alkalinity for smooth operation. Trisodium phosphate is therefore added to the drum directly for solid alkalisation of drum water for continuously maintaining an optimum pH. Other solid alkalis including NaOH, etc. could also be used, but as it will be seen later, trisodium phosphate has been seen to be the best option. Under suitable dose, this also helps in maintenance of protective layer of magnetite on boiler metallic surfaces. Though this phosphate is liable to form sludges with calcium & magnesium salts coming to the boiler through condenser leaks, etc. modern boiler operations do not allow any solid loading to cause precipitation. Thus, phosphate dosing is required to achieve only due pH, for which a minimum possible amount needs to be dosed.

Conventional Phospate Treatment

Adopted in 1920, it was used not only for solid alkalisation of boiler water, but also to prevent calcium carbonate scale in the boiler in the event of condenser leak if any, in which way calcium is precipitated as calcium phosohate giving soft sludge to be removed by boiler blowdown and similarly magnesium is also precipitated & removed as magnesium silicate in presence of silica, otherwise as magnesium hydroxide. But in this way, molar ratio of sodium & phosphate Na / PO was needed to be maintained greater than 3.0 and pH 11 to 12. But in this way, it was found that high alkalinity maintenance was essential which gave non-adherent particles of calcium phosphate and avoided formation of adherent magnesium phosphate.

Magnesium hydroxide or silicate as precipitated will also neutralise acid chloride if found in the boiler water or else removed by blowdown. This type of dosing however could not be found suitable for a high pressure boiler in view of sludge formation and under deposit corrosion by formation of free sodium hydroxide as per following reaction. Oil & organics content were also found to give adherent deposits.

$$Na_3PO_4 + H_2O \rightarrow Na_2HPO_4 + NaOH$$

Formation of more acidic phosphate in the boiler water as formed under following reaction, is also undesirable, which is not useful at all in protecting against corrosion, but rather this will overload the boiler with sodium for carryover to steam.

$$Na_3PO_4 \rightarrow Na_2HPO_4 \rightarrow NaH_2PO_4$$

Coordinated phosphate / pH control

The practice adopted in 1940 is to have pH buffer in boiler water to limit the localised concentratition of caustic soda. Control was achieved through feed of the proper type of phosphate to either raise or lower the pH with maintenance of proper phosphate level by blowdown, as per the following vector control diagram;

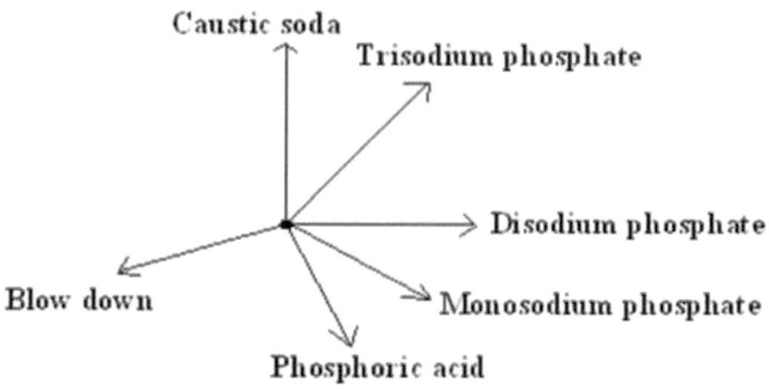

This gave different control area with different operating pressures of boilers, viz.,

Control area	Molar ratio Na/PO_4	pH
< 900 psi	Captive Alkalinity Region	9.8 - 10.7
901 - 1500 psi	2.8	9.3 - 10.3
1501 - 2000 psi	2.7	8.9 - 9.8
2001 - 2500 psi	2.6 | Free Caustic	8.7 - 9.4
> 2600 psi	2.6 | Region	8.4 - 9.1

Various combinations of disodium phosphate and monosodium phosphate injections and feed rates, blowdown adjustment and caustic addition are used to maintain proper phosphate/pH levels, limiting the maximum boundary of Na/PO_4 = 3.0 and minimum of 2.2. This, however, gave more serious problems of caustic gauging corrosion. Phosphate was found to react with corrosion products to form various iron complexes and even the protective magnetite and the liberated NaOH remaining underneath complex. On ratio going below 2.1, H_3PO_4 formation was also noted. A common iron complex called Maricite ($Na\ Fe_3PO_4$) was formed when magnetite was reacted with either disodium phosphate or monosodium phosphate.

Congruent Control (CPT)

Attempt was therefore made in 1954 to dose phosphate in such a way that 'hideout' phenomenon of phosphate from the system is not by reacting with corrosion products, but only through congruent precipitation of pure Na_3PO_4 (Molar ratio Na/PO = 3.0) and affecting no pH change with phosphate dosing as shown in the following diagram

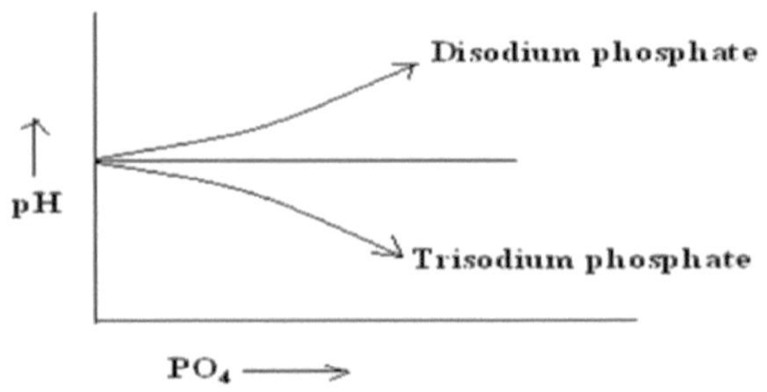

This was found possible only when impurities or corrosion products are removed immediately from the system & very low dosing of phosphate in a balanced way was done.

But even pure Na_3PO_4 (Na/PO_4 = 3.0) was found able to produce free NaOH after supersaturation and there was formation of solid compound having Na/PO_4 = 2.85. This required boiler water pH control = 2.85 ideal below Na/PO_4 being at 2.8.

Though the ratio maintenance at 2, is continuously difficult, but has been achieved by continuous feed of phosphate and continuous blowdown. It was felt that if caustic was to be avoided to avert localised under deposit corrosion, the pH of boiler water should be maintained below the level corresponding to sodium phosphate solution having a ratio of Na/PO_4 of approx. 2.6 during which boiler water pH would remain stable despite local saturation and solid precipitation.

Equivalent Phosphate Treatment (EPT)

Even with congruent treatment, when few boilers faced acid phosphate corrosion, mostly with hideout phenomenon, and with trip induced chemistry resulting to drastic pH drop of boiler water, addition of more phosphate became more harmful. To prevent utilities from having to chase phosphate hideout and associated addition of any form of phosphate, phosphate treatment was followed to maintain Na/PO_4 = 2.8 plus 1 ppm NaOH.

This treatment was called equivalent phosphate treatment (EPT), which becomes 1 ppm excess NaOH above coordinated pH/ PO_4 curve. Thus, phosphate treatment (PT) becomes an extension of EPT at higher levels of phosohate or within 2 - 5 ppm PO_4 limits of CPT (pH 9.0 - 9.5), as shown in following figure.

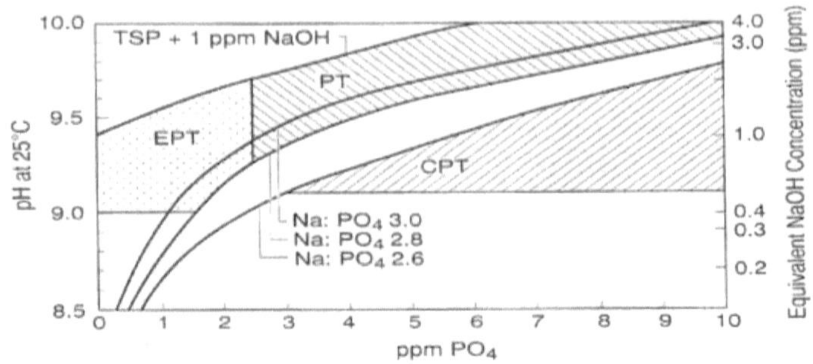

Schematic of operating ranges of boiler water on EPT, CPT and PT, an EPRI licenced material.

EPT is thus the best solution if trip induced chemistry change is found with significant excursions in boiler water's pH, but under EPT, it is necessary that level of maximum phosphate concentration is decided by test on chemically cleaned boiler and operated for a week without chemical control & blowdown and most importantly without any contaminants, which is not guaranteed. Even if this is made possible, frequent changes in EPT level are there required for reassessment. Thus, risk of caustic under deposit corrosion, hydrogen damage and caustic gauging continue to persist in high heat flux boiler.

Equilibrium level of phosphate in an operating boiler is determined as below:

When the unit is seen to be operating over a week without any control or contaminants/problems and with the blowdown closed, perform this test by reducing unit load to a minimum value and keeping the boiler blowdown closed. The phosphate concentration is adjusted using tri sodium phosphate only, to near the upper limit with pH 9.5 - 9.6 (corrected for ammonia) and with not more than 1 ppm free NaOH. Discontinue phosphate feed & bring unit to full load. After several hours, phosphate is reduced by hideout finally reaching a minimum level, which is the equivalent phosphate concentration and operate later only up to this level of phosphate. However, during test, particular attention has to be paid for sodium level in steam.

Preferential Choice of Phosphate Treatment

There is therefore need to evaluate problems on site assessment and type of phosphate dosing fixed accordingly. Operational procedure, burner configuration, heat flux, deposit density, fuel changes, firing rate, boiler cleanliness, etc. all play roles in the amount of phosphate that a boiler can tolerate. So far, congruent control of phosphate in most cases have been found the most suitable choice, keeping in view maximum phosphate concentration, a boiler can tolerate with minimum hideout of phosphate. This maximum phosphate concentration should be close to equilibrium concentration i.e. defined as concentration tolerated at maximum load without forming solid phases on heat transfer surfaces. Thus, usual level of phosphate content in boiler water should be as low as possible (typically in the range 1- 2 ppm) only to attain optimum pH, in which adequate contribution of ammonia dose has also to be taken.

Over and above, if in any utility, the feed water quality is being regularly able to achieve as Fe less than 5 ppb and Cu less than 2 ppb, no change in existing phosphate treatment is required.

Treatment Other Than Phosphate

Sodium Hydroxide:

This has been in practice in CEGB (UK) power stations. Under the practice, it is necessary to maintain sodium hydroxide and sodium chloride level proportionally with online monitoring and automatic dosing & blowing down the boiler facilities to avoid any lag behind in controlling the due proportion of hydroxide & chloride. By this way, not only due alkalinity to boiler water is maintained, but the practice has been found more effective in combating the acidic corrosion, that is likely to occur in the boiler in case of leakage of cooling water, which in most cases have been sea/brackish water in UK.

However, sodium hydroxide has the disadvantage that if its concentration is not effectively controlled, it may get concentrated under the deposits to lead to the caustic embrittlement at the site. The effective control is also not possible manually and also in the presence of impurities like corrosion products, etc. which are always apprehended in the boiler having frequent shutdown & start ups. On NaOH treatment practice, boiler water quality has to be very precise to get steam of required stand and in respect of sodium content and cation conductivity, in view of the usual mechanical carryover from boiler to steam is more than 0.1%. NaOH treatment practice is also not advisable on the boiler having its make-up water requirement greater than 1.0% and on the boiler having multiple high heat flux zone.

All Volatile Treatment

In this treatment, there is no chemical dosing directly to drum. Neutralising & filming amines along with ammonia is added to boiler feed water, in addition to hydrazine as oxygen scavenger. But its satisfactory working in drum type boiler with mixed metallurgy in feed system can be guaranteed only on clean system and on base load operation, and that also with very controlled dosing of the chemicals, which is however difficult for a unit having frequent trip & restart and having corrosion products and other impurities. Simple ammonia/ volatile amine generally fails to have corrosion protection in case of upset of boiler water quality. Vaporous carryover of salts from boiler to steam in presence of these, is also increased which may lead to turbine blade corrosion & disc cracking. AVT has however been found suitable for once through boiler and where condensate polishing are being done satisfactorily.

AVT undoubtedly requires a very clean condition of operating boiler with no tolerance of even little corrosion products, otherwise more serious situation for instigating corrosion could be created. The boiler water cation conductivity should always be below 1.5 micromhos/cm, Fe content less than 10 ppb and dissolved oxygen 5 – 10 ppb. Hence, towards this there is need of 100% condensate polishing, if AVT is a preferential choice.

Hydrogen Peroxide/Oxygen Treatment (OT)

Recent trend though universally not practiced so far, is to add H_2O_2 in the form of dosing limited amount of oxygen to the boiler water. Protective magnetic oxide is formed here as Haematite (Fe_2O_3), in place of Magnetite (Fe_3O_4), which give more resistance to corrosion. The system should however be all ferrous and the principle lies in neutral water treatment at pH~7.5 together with addition of H_2O_2 or gaseous oxygen. When adding both H_2O_2 & oxygen, the pH must be maintained at no less than 7.0, for which help of small quantity of ammonia dose is also taken so that pH remains in the range of 7.0 - 7.5 only. Optimum H_2O_2 concentration was established to be of 500 - 700 microgm/kg for minimum corrosion. On decomposition, H_2O_2 gives OH plus OH radical. The latter gives hydroperoxide complex of iron, converting to protective magnetite film on condensate and feed line at 140° C and later haematite protective film in the boiler region. Hence, it is preferable to add H_2O_2 to the suction of condensate pumps. Oxygenated treatment however requires very pure feed water of cation conductivity of less than 0.15 micromhos/cm.

In spite of the above chemicals found sufficiently competent to condition the boiler feed and boiler water itself, surveys indicate that half of the TPS maintenance costs and half of forced outages are attributable to corrosion only. Water formed deposits as found on earlier days are mostly over, in view of use of ultra pure quality of water as make-up now. The deposits, if any, found now are of mostly corrosion products associated with the contamination of impurities as found in the chemicals in use for conditioning and impurities carried over from the make-up water/DM plant.

The nature of deposits

As generally found in the boiler is thus mainly iron oxides contaminated with copper oxides (if copper metallurgy is also employed in the system) and sodium salts, viz., phosphate, chloride, hydroxides, etc. along with silica & some organics contributed through either chemicals or make-up water. Such deposits if found in considerable amount in the boiler, means possibilities of these contaminants having carried over to steam during actual operation to deposit on the turbine blades, SiO_2 deposits on turbine generally give a very hard incrustation on turbine blades very difficult to remove and could be found in different phase composition on the blades of high pressure turbines as below:

a.	Quartz	(SiO_2)	formation range	94 - 288 ^{0}C
b.	Amorphous silica	(SiO_2)	formation range	49 - 255 ^{0}C
c.	Cristobalite	(SiO_2)	formation range	79 - 182 ^{0}C
d.	Sodium metasilicate	(Na_2SiO_3)	formation range	280 - 372 ^{0}C
e.	Sodium metabisilicate	(Na_2SiO_5)	formation range	216 - 377 ^{0}C
f.	Acmite	($Na_2OFe_2O_3$ $4SiO_2$)	formation range	210 - 395 ^{0}C

An increase in the alkalinity of the boiler water led to a decrease in the distribution coefficient of silicic acid between steam & water. However, when sodium meta silicate is substituted, the SiO concentration in steam gets stabilised, as after equilibrium, hydrolysis of sodium metasilicate to H_2SiO_3 is reduced. When the NaOH: SiO_2 ratio in boiler water was kept at around 0.25, a predominant deposition of silica on turbine blades occurred. When this ratio was increased to 5 - 10, the deposits then consisted mainly of sodium bisilicate, which is water soluble. With the reaction of NaOH with iron oxides and silicic acids, 'Acmite' sparingly soluble in water is formed as turbine deposits.

With steam expansion, impurities present in steam start separating out & deposit on turbine, since impurities solubilities in dry steam decreases as the steam expands. The degree of wetness of seam leaving last row blade should not be more than 10%. At the same time, it must not be supersaturated also. Expansion rates of steam remaining as high as 3000:1 in one second, it takes only about less than 0.2 seconds for steam to expand from HP inlet to the condenser. Yet, even the smallest concentration of impurities is causing some times severe turbine corrosion problems and deposits. For example, 10 ppb of NaOH in steam is sufficient for formation of 80% solution of NaOH in IP stage and such separation can lead to stress corrosion cracking. Similarly, chloride, organics and silica in steam need to be limited to 5, 100 and 20 ppb respectively.

Potential effect of organics decomposition on stress corrosion or cracking of turbine components like 'Disc' have also been reported. However, for all types of turbines, the same contamination limits cannot be followed, but since the turbine is the most sensitive component for many of the contaminants, viz., sodium, chloride, silica, corrosion products, organic carbon, phosphate, sulphate, oxygen & pH (alkalinity/ acidity), it is the steam chemistry which sets pace for working backward for the possible chemistry and qualities of boiler water, feed water, make-up water, etc. in series to be tuned up in respective order.

Turbine Steam Chemistry

Its chemistry mainly involves with corrosion and erosion and is studied along with superheated steam section, which is generally of alloy steel/austenitic steel construction and operating at 520 to 550 ^{0}C and at a very high pressure up to greater than 165 kg/cm^2 (for 500 MW & above). Causes of corrosion in these sections are both salt induced and deposits and impart mainly local corrosion & stress corrosion and corrosion cracking of turbine blades & discs. While SiO_2 & metal oxides cause local corrosion as deposits induced, NaOH & NaCl mainly give stress corrosion. NaOH also cause local corrosion by way of pitting & stress. Sodium, if found in the form of Na_3PO_4 is generally non-corrosive, but may cause under deposit corrosion. The deposit is, however, mostly washed away during shutdown period of turbine. Yet, out of 5 ppb of sodium in steam, 2 ppb can be assumed to be associated with phosphate & 3 ppb as NaOH & NaCl, if conventional phosphate dosing (as later described under 'boiler water chemistry') is in practice. For boilers operating on EPT, all sodium can be taken in NaOH form. In steam, to avoid deposit of NaOH on turbine, sodium should however be preferably limited to 3 - 5 ppb as Na_2CO_3, if found in the steam is responsible for corrosion at wet end of turbine and organic acids for turbine disc cracking. A typical deposit of 13% NaOH and 28% Na_2CO_3 has been found highly corrosive to turbine steel. Steam velocity and turbine blades design factors have often been found to cause erosion of turbine blades. Corrosion fatigue & stress corrosion cracking of turbine blades and discs are more commonly associated with even traces amount of S, Cl, Na & organics, and problems have been more common in low pressure part of turbine which has high stress, crevices and temperature conducive to the condensation of concentrated solution of steam contaminants. If steam is found more impure, this can occur in high pressure part of turbine also.

In wet steam region, where temperature of metallic surfaces exceeds that of steam, evaporation & drying cause impurity concentration & corrosion. The moisture may also cause a surface film of a concentrated solution. The boundary of a dry area and surfaces experiencing frequent change of conditions can be particularly prone to corrosion. Impurity concentration is more rapid in areas supplied with fresh moisture, as opposed to stagnant areas. Use of film forming amines are recommended at this stage like cyclohexylamine/morpholine, etc. into low pressure (LP) turbine steam, in view of their less distribution ratio in steam, but since these amines are generally organic in nature, they create more CO including organics atmosphere in feed water-steam cycle. Also, stress corrosion cracking of some turbine components including turbine discs due to more total organic carbon presence in steam cannot be overruled. Due to organics, there is considerable increase in cation conductivity also in boiler feed-steam cycle.

There is another way to combat this acidic region of turbine. In some utilities, to increase both mixing and reaction time with oxygen, hydrazine is sometimes injected into the IP-LP turbine crossover pipe. This injection also increases the in situ pH at turbine surfaces near the saturation line, helping to neutralise acids. The distribution ratio of hydrazine in steam to moisture in these steam conditions is about 0.03 at the recommended level of 20 ppb of hydrazine in steam, there will be about 700 ppb of hydrazine in moisture resulting in a pH increase of approx. 0.5. In view of hydrazine being carcinogenic and its use being discouraged, use of ammonia alone is made in spite of its distribution ratio in steam being higher. But having all ferrous system metallurgy at 'pre-boiler system' at steam temperatures, there are also chances of hydrazine decomposition in steam into hydrogen and nitrogen.

Undesirable stresses and impurity concentrations can also be caused due to turbine design, its metallurgy, physical shape & surface finish also, which can cause crevices, stress residual, obstacle to flow with respect to moisture velocity, location of salt zone, stagnation temperature, interaction of shock wave with Wilson line, etc. Some turbine manufacturers like to choose lower yield strength material in turbine blades & discs to mitigate the problem of more impure steam problems, but in large size units, impurities in steam cannot be tolerated at all. Hence, it is necessary that boiler design and its operation should be conducive to obtain a high purity steam for rolling the turbine. Salts & impurities entering the steam has to be stopped, which are generally due to two main reasons:

a. Mechanical carryover of salts through water droplets, for which drum internals have to be more accurate to arrest water droplets in such a fashion that efficiency of separation of steam from water is better than 0.2%. The total carryover of salts in steam thus should not be more than few parts per billion, before using it to rotate the turbine.

The following are generally the safe operating limits of turbine steam purity:

Sodium as Na, ppb	5 - 10
Silica as SiO_2, ppb	10 - 20
Chloride as Cl, ppb	< 3
Sulphate as SO_4 ppb	< 3
TOC as C ppb	< 100
Cat. Cond. µS/cm@25°C	< 0.3

b. Working backwards and based on 0.1% mechanical carryover from the boiler (drum type), the following boiler water chemistry along with its impurity limits need to be followed, which are generally pressure dependent of the operating boiler: Whereas, at 1200 psi, mechanical carryover should be lesser than 0.1%, the same at 2650 psi should be 0.2% i.e. higher is mechanical carryover with higher drum pressure.

Deposits on turbine cause more consumption of coal and lower the efficiency. However, a 500 MW unit can tolerate more deposits without this detectable effect.

Boiler Water Quality

Na, Cl, SO_4 & SiO_2	As per pressure curves
PO_4 & pH	As per PO_4 - pH curve
Na/PO_4 molar ratio	2.6 - 2.8

The pressure curves provide opportunity of relaxing boiler water quality particularly during transient chemistry for short period and remove over concentration of salts by reducing pressure. Therefore, drum type boilers generally have capability of sliding pressure to decide right action level.

Immediate shutdown of boiler is required at and below pH of 8.0.

It is not sufficient just to maintain the concentration of the impurities in the boiler water, but it also becomes necessary to add some chemicals to it along with giving blowdown, to ensure adequate pH with the impurities transported to it through feed water. In most of the boilers, the drum water gets concentrated 10 to 20 times of the boiler feed water and hence,` it is necessary that a good quality feed water is envisaged for the boiler. In other words, boiler water chemistry will no doubt be mostly dependent on feed water quality, but corrosion and magnetite protective layer solubility of the boiler surfaces have to be kept under control to avoid contribution of impurities from boiler parts itself.

Boiler Water Chemistry

Quality of boiler water should be such that it not only assists in the formation of magnetite layer over all the tubes and drum surfaces, but also maintain a thin adherent film all along & all through operation and during its layout also. With this aim, concerned impurities which impair this passive protective layer, must be removed from the boiler tubes & drum. A look at the potential pH diagram in respect of mild steel corrosion and magnetite layer solubility at 300 ^{0}C (as shown in fig below) show that in order to prevent both the corrosion & magnetite solubility together, the best optimum pH is 9.25 (25 ^{0}C). Any reduction in alkalinity is only possible when chloride, dissolved oxygen, sodium hydroxide and other impurities including corrosion products are minimised, otherwise passive oxide film is likely to break, leading to progress of corrosion. Thus, concentration of impurities mainly acidic or alkali components in excess has been known to be prime factors for onloadcorrosion, which need to be avoided. Hence, in once through boiler, no non-volatile chemicals are used for boiler water pH control, but in drum type this is necessiated for the reasons of high distribution of these volatile chemicals. Other causes of onload corrosion to boiler are:

a. Aggressive contaminants in boiler, viz., chloride, oxygen, etc. Chloride together with Ni, Co, V, Sb would not allow to form protective magnetite and cause rapid corrosion.

b. Local concentration of dilute solutions within boiler, like (i) steam separation at heat transfer surfaces leaving behind solute at crevices, welds, joints, bends, places of deposits (of iron and copper oxides) causing 'wick boiling', i.e. boiling without mixing with the bulk solution because the liquid is fed to the surface through small pores in the deposits (porous in nature) like a 'wick' or flow disturbances at bends or welds.

c. Steam blanket at the heat transfer surfaces in the top half of horizontal or inclined tubes heated from above. Solute concentrate when the dry surfaces are intermittently wetted or splashed.

d. Hydrogen generated in the system by reaction of metallic iron with water steam, $3\ Fe + 4\ H_2O \rightarrow Fe_3O_4 + H_2$; or by any means like decomposition products of chemicals, reacts with carbon or carbide of steel called 'decarburisation'.

In offload condition, the oxide formed on low alloy steel surfaces is generally porous and does not prevent further corrosion (as found in metallic iron, aluminium, chromium, etc. forming passive layer, preventing further corrosion), but in fact accelerate the rate of attack. The corrosion is also caused when water becomes trapped at poorly drained sites e.g. tube sags after blowdown, in economisers, etc. During offload, unlimited access of air above water provides and Oxygen, which carbon dioxide stimulates corrosion.

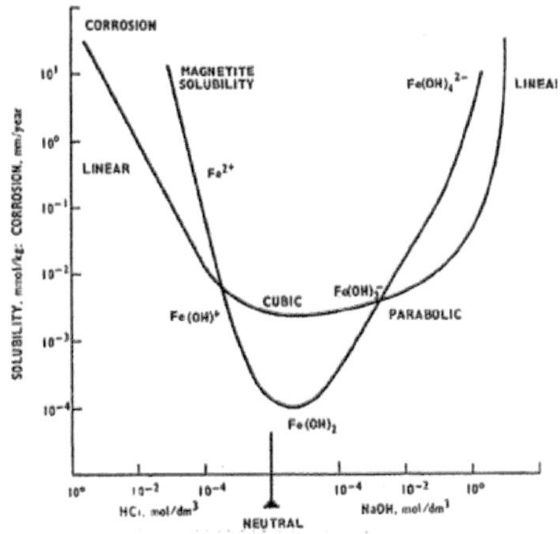

Corrosion of Mild steel & Solubility of Magnetite at 300 ^0C (A CEGB Bulletin under its training programme)

The current practices of chemical dosing to boiler to avoid corrosion on its surfaces are with any one of the following:

a. Sodium hydroxide treatment
b. Sodium phosphate treatment
c. All volatile treatment including ammonia, neutralising amines, etc. d
d. Hydrogen peroxide (oxygen addition) treatment. Each of the above treatment has its own advantages & disadvantages, which are discussed in detail as below:

Feed Water Chemistry

Working further backward to boiler water, maintenance of chemistry of feed water to boiler is very important in view of relative surface area being much greater than the boiler regime and no protective magnetite layer formation possible in these areas due to being lower temperature region. There are not only chances of more corrosion products, but presence of other metallic oxides other than iron, also cannot be ruled out in these regions in view of different metallurgy if used for different equipments like condenser tubes, heaters, etc. All these, along with other impurities are likely to be transported to disturb the boiler water chemistry, if proper feed water chemistry is maintained. The following could be seen as the root causes of corrosion in these areas i.e. pre-boiler, which comprises condenser & hot well, deaerator, LP & high pressure (HP) heaters, economiser, etc.

Oxygen:

Sources of these are make-up water. More the make-up, more is dissolved oxgen (DO) and air in-leakage to condenser. This reacts with metallic constructions iron or copper like this:

$4\ Fe + 6\ HOH + 3O_2 \rightarrow 2Fe_2O_3 \rightarrow 4\ Fe(OH)_3$ (reaction goes on continuously at measurable rate & doubles at every 30 deg C temperature rise)

$2\ Cu + O2 \rightarrow 2CuO$ (reaction self limiting type and stops after forming film)

Oxygen is known to be highly corrosive in presence of carbon dioxide, ammonia, sodium chloride & sulphate. It reacts with chloride synergistically, resulting in hydrogen damage. Further, for corrosion reaction, it is rate of arrival of oxygen on the metal surface which is more important than the concentration itself.

Carbon Dioxide:

Sources of this are make-up water, air in-leakage to condenser and decomposition of organic matters finding entry to feed water through either make-up or chemicals feed to it. The reaction goes like this:

$2\,H_2CO_3 + Fe \rightarrow Fe\,(HCO_3)_2 + H_2$ Hydroxide and oxides of Fe (reaction is of limiting type, but free CO 0.5 ppm and above can accelerate the attack of corrosion)

$$2\,Cu + H_2O + CO_2 + O_2 \rightarrow CuCO_3\,Cu(OH)_2 \rightarrow CuO$$

Ammonia / Amines / Hydrazine:

Sources of this are excess dosing of ammonia/amine (usually above 1 ppm) and also of hydrazine (residual of more than 0.05 ppm; any excess is liable to be decomposed to ammonia in boiler for recycling to feed). Ammonia/amines do not generally react with steel surfaces until they furnish a very high pH (10.5 and above & not in practice in TPS), but ammonia neutralises the acidic Fe (OH) and in the process generate ferrous ammonium carbonate complexes of unstable nature, known to have interfered with cation conductivity. Thus, in operational practice, ammonia/amine in excess is likely to give adverse effect on copper metallurgy at large, as below; in presence of oxygen. However, even low ammonia is detrimental to copper.

$$Cu + NH_3 + H_2O + O_2 \rightarrow Cu(NH_3)_4^{2+} + 2OH^-$$

The complex is generally formed from the copper oxides film on the tube surface & the decomposed CuO on heat is carried over with water flow. With amines, reaction is however slow, but forms more stable complex of copper of varying nature. Apart from the above corredents, viz., oxygen, carbon dioxide, ammonia/amines, etc. it has to be remembered that presence of corrosion products particles itself is a nucleus to accelerate corrosion and this must be removed immediately from the system.

Many utilities have found that addition of hydrazine in all ferrous systems make very little difference at the levels of oxygen being maintained at the deaerator outlet (<5 ppb) and hence have been able to eliminate its use. Reaction kinetics of hydrazine with the dissolved oxygen at such level is very poor. Rather, very low levels of oxygen (much lower than 5 ppb & to the tune of 1 ppb) in conjunction with high levels of hydrazine (much greater than 5 ppb to the tune of 10 ppb) result in reducing condition to generate iron oxides in feed water system.

Cycles with copper alloy in feed water system may however require lower oxygen levels and higher hydrazine levels than cycles with all ferrous components. Copper acts as catalyst to improve the reaction kinetics of hydrazine and oxygen. Further, since corrosion products have also been identified to have catalytic effect on these reactions, maintenance of 5 ppb of oxygen and less than 20 ppb of hydrazine in feed water have been taken as ideal in most of the TPS having often transient chemistry.

Overfeeding of hydrazine is however not recommended, which may accelerate erosion-corrosion of carbon steel, and a few ppb of oxygen in high purity water may be beneficial in reducing generation of iron corrosion products. Where there is condensate polising unit, higher doses of hydrazine may create hydrazine sulphonate in the resin, which is hard to be removed.

Further, to minimise pre-boiler corrosion, an understanding of the operational requiremens of condenser hot well, feed water heaters (both LP & HP), deaerators, economisers & other critical system components is necessary. The primary problems in heaters are corrosion due to oxygen & improper pH and erosion from the tube side or the shell side. Due to temperature increase across the heater, incoming metal oxides are deposited in the heater and then released during steam load change & chemical imbalances. Stress cracking of welded components can also be a problem. Erosion is common in the shell side, due to high velocity & steam impingement on tubes & baffles. Condenser hot well is also prone to corrosion due to high boiler make-up water requirement, temperature variations & non-removal of adequate dissolved oxygen from water and air in-leakage to condenser, etc. Condenser tubes also have some direct impingement of wet steam to cause thinning of metal, apart from incondensible gas attack. In deaerators, corrosion fatigue near welds, stress corrosion cracking of stainless steel tray chamber, oxygen pitting of vent condensers & erosion of impingement baffles near steam inlet connection, etc. are common. Oxygen pitting caused by presence of oxygen & temperature increase is major problem in economisers. Steaming economisers are very sensitive to deposition from feed water contaminants & resultant under deposit corrosion. Erosion at tube bends is also a problem in steaming economisers.

Resin fines carried over from water treatment plant to condenser hot well or styrene sulphonic acids released from oxidation of strong acid cation resin in condensate polisher (which interferes in determination of sulphate by IC) can also potentially contribute as is or as their pyrolysis products to corrosion products generation and transport to boiler.

Thus, to achieve a good behaviour on the part of boiler, it is necessary to pay attention to secure the following type of quality of feed water all along its operation:

Iron, as Fe	ppb	< 10
Copper, as Cu	ppb	< 2
Oxygen, as O_2	ppb	< 5
Hydrazine, as N_2H4	ppb	< 20
pH @ 25 ^{0}C (Cu alloy)		8.8 - 9.2
pH @ 25 ^{0}C (ferrous)		9.0 - 9.6
Cation conduct. μS/cm@25 ^{0}C		< 0.3
Ammonia as NH_3 ppm		consistent with pH

In view of cation conductivity being the ready measure for the purity of feed water– steam cycle in a TPS, the following table in the form of fish bone diagram can be found helpful in finding out the cause of deviations, if any, to approach suitable actions accordingly:

Key Approaches on Operational Part of a Boiler

The following points to be looked into by boiler operating staff, water chemistry personnels also need to be aware of, for suitable actions accordingly for a good water chemistry control of boiler water for trouble free operation of boiler:

a. The basic function of a steam drum is to provide a large volume, to reduce the upward velocity of the steam/water mixture, because the moisture droplets are denser than steam, they fall back to the water surface. Drum internals e.g. baffles, plates, screens & centrifugal separators help reduction in drum size requirements & provide a means for elimination of the finely divided moisture spray, which otherwise could not be removed.

The screen, drain pipes, hence, must terminate below the water level of the drum, so that possible moisture bypassing of dryer screens are avoided. Also, feed water inlet nozzles must be aimed below horizontal to eliminate direct entrainment of feed water.

b. An understanding of boiler circulation dynamics is helpful in tuning 'Drum-level' control. In this regard, therefore, feed water flow should be proportional to steam flow with reset action on drum level. This means that pounds of water entering the drum should always equal pounds of steam leaving. Also, feed water supply capability should never be completely shut off, not even for a short period of time."

c. Each off-on cycle of a unit on cold start is equivalent relative to the fouling of heat transfer surface to one month of continuous operation unless special condensate & feed water flushing/polishing steps are taken.

d. When boiler is being filled with hot water, it should be seen that the feed must be slow as per its operating condition to avoid severe temperature induced stresses. Filling hot boiler with cold water must also be done very slowly. Also, during filling and initial firing, the drum vent must remain open until all air is vented from the unit and an appreciable amount of steam is issuing there from. The unit may be considered free of air when the drum pressure reaches 25 psi.

e. Drainable portions of superheaters (headers, inlets, outlet & connecting piping) should be drained prior to lighting off through lines free from any back pressure and the outlet should be vented. The superheater drain should then be closed, with the exception of the superheater outlet vent (drain), which must remain open to ensure a flow of steam to protect the superheater elements from overheating until boiler has gone online and is carrying load. The vent or drain may be throttled as the boiler pressure increases.

f. During initial start up, the unit should be kept at a minimum firing rate for as long as necessary for checking proper functions of various equipments/ instruments. The low firing rate must be maintained until the boiler is carrying load. Overfiring the boiler for any reason should be avoided, even it to match turbine metal temperature during hot restart up. Pressure can be raised at a rate corresponding with the start up curve, which allows generally 3.5 to 4.5 hours for large moderate pressures boilers having superheaters.

g. Boiler operation is more critical during the light off period than at any other time. It is during this time that the boiler can be damaged by overfiring, failure to purge, failure to protect the superheater, furnace puffs, etc. While shutting down the boiler, reduction of saturation temperature should be approx. by 1000 F per hour and at 25 psi. Drum vent should be opened (if not opened earlier). The boiler, if it has to be drained, should be drained hot invariably above 25 psi, to keep it moisture free during its idle period.

h. Sufficient flow through economiser needs to be maintained to prevent the generation of steam in the economiser, which is generally characterised by a rapid rise in steam drum water level and/or water hammering when feeding water.

i. Use of bottom blowdown valves should be restricted to periods of moderate steaming rates and at preferably very low combustion rate, and never to be operated at high loads. Blowdown through these valves have to be done to free a boiler from sediment, when the boiler is being brought off the line. The valve closest to the boiler needs to be opened first than the valve farthest from the boiler. After execution of blowdown they have to be closed in reverse order. The blowdown have never to be done to lose sight of of the water in the gauge glass.

j. For boilers operating above 350 - 400 psi, most gauge glasses utilize mica inserts, which deteriorate rapidly, when draining of the gauge/or blowdown are made very frequently or with wide open valves slowly opening the blowdown valve is usually sufficient to check the level on these glasses.

k. A sudden drop in steam temperature is indicative of excess carryover of boiler water from drum to superheater caused by high drum level, foam formation or drum internals malfunctioning. The proper response is to decrease load & take suitable action by blowdown, shutdown & flushing of steam lines.

Abnormal superheat steam temperature (which is generally fixed with different loads) can be read on the following lines:

	Abnormally High Superheat Steam Temperature	Superheat Steam Temperature below Normal for the Load
1	Too high excess air	Too low excess air
2	Too low feed water temperature	Too high feed water temperature
3	Heavily slagged furnace	Excessive moisture carryover the boiler into the superheater
4	Secondary combustion	Excessive external deposits on superheater tube
5	Improper operation of steam temp. indicating and/or control equipment	Improper operation of steam temp indicating and/or control equipment

l. Filming amines can slough condenser surfaces & deposit in boiler when overfed. Based on amine, special boiler chemical cleaning programme has to be initiated.

m. Flue/stack gas temperature may be read for its abnormalities as below:

Below normal	Above normal
Very low excess air	High excess air
Very high CO_2	Fouled fire sides
Some CO	Secondary combustion
Smoke	Leaky baffles A fire in the air heater

n. Above all, the integerity of the entire steam-water circuit need to be maintained to limit ingress of impurities in the system by keeping attention on the following:

1. Good make-up water supply generally of ultra pure nature.
2. % make-up water requirement optimisation to +/- 1% of design value.

3. Proper hydrazine/ammonia dosing in feed water & phosphate in drum.
4. Control of DO, pH, Cu & Fe in feed & Na/PO in boiler water and also ammonia in condensate.
5. **Close watch on:**
   a. Drum level fluctuation to limit within 0+/-150 mm.
   b. Attemperation rate to limit to minimum.
   c. Air in-leakage to condenser; dissolved oxygen & condenser vacuum to be maintained as per design DO at condenser should not be greater than 0.015 ppm.
   d. Condenser leak to keep watch on Na, Cl & cation conductivity variations.
   e. Operational procedures optimisation in the event of

Condenser leak	Cold start up
Load fluctuations	Other transient state
Boiler preservation	Boiler Cleaning
Condensate polishing	Stator water polishing

   f. Boiler heat flux has to be limited, when boiler water is dirty. During this time blowdown needs to be maximised and firing rate restricted to not more than 50% power at full pressure as long as boiler water suspended solids are present. After that, if found necessary, boiler pressure can be brought up as silica limitation permit, viz., there is not more than 0.02 ppm of silica in steam at any pressure of boiler.
   g. During steam blowing (commissioning period), the boiler water should be chloride free to avoid stress corrosion & only be treated with volatile chemicals with reserve of ammonia only as 10 ppm. This will also avoid the risk of mechanical carryover due to more operating pressure drop.
   h. Concentration should decrease as pressure (load) decreases, due to change in steam/water ratio in the boiler, i.e. higher % of water is now available. More concentration means 'hideout' of salts taking place.

i. It has to be kept in mind that leak developed in LP heaters may help water get into turbine.

j. Superheated steam temperature will drop suddenly and recover, if the boiler is priming, i.e. total dissolved salts or alkalinity increase.

k. To avoid the concentration of high amounts of ammonia in the system, the after air ejector drains should be continuously wasted.

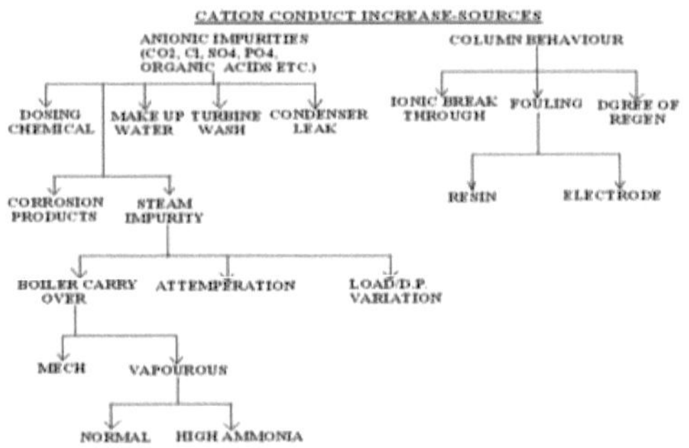

Apart from paying attention towards maintenance of quality parametres and operational practices accordingly, there is much more to be looked into, for safe operation of boiler and turbine, during transient chemistry occurence, like condenser leak, offload preservation of boiler, hydrostatic test, etc. together with proper functioning of condensate & stator water polishing units, along with the sampling points, DM water storage tank, etc. and chemical cleaning of boiler (if required) which may have direct impact on the performance of boiler water or turbine steam chemistry. General guidelines towards this are as follows:

Healthy Operational Practices

1. In the Event of Condenser Leak

When the condenser leak is proved by the analytical data of cation conductivity, sodium and chloride content of condensate, the following steps need to be taken:

a. Reduce the load till attemperator water is not required. If relevant, make-up water to stator coolant to be caesed from condensate and to be given from DM water.

b. Continuous blowdown valve of the boiler has to be kept fully opened and allow a very good make-up water from DM plant duly chemical dosed. Chemical dose to boiler water has also to be increased to maintain the reserve of alkalinity and keeping sodium, chloride, etc. under control. If the unit is running on AVT, caustic soda can be added to maintain up to 3 ppm NaOH in boiler water.

c. If relevant, put condensate polishing plant into service.

d. Plan to check the leaking tubes and have leak attended at reduced load.

e. If it is not possible to maintain the boiler water pH into limit in spite of due chemical dose & blowdown, stop the boiler to attend the leak.

f. On sea water cooling and having no condensate polisher, the unit should not be tried to run any longer.

g. After leak attending check the cathodic protection to the tubes are intact, and then drain & flush the condenser & feed system suitably before bringing the boiler up again.

2. **Offload Presevation of Boilers:**

Moisture, oxygen, low pH conditions & some salts present in boiler water enhance corrosion during offload and thus three types of storage conditions are practiced to counter act these factors, as below:

a) Dry storage in which moisture is eliminated by use of hot/dry air or while offloading the boiler, draining the system rapidly under pressure & temperature and boxed up. Recommended practice for long outage or whenever boiler needed maintenance and internal inspection.

b) Elimination of oxygen by use of nitrogen but having the plant wet including use of hydrazine in adequate amount to react with dissolved oxygen of water. After draining of the plant, nitrogen along with ammonia gas is passed into the system to displace the air & boxed up under pressure.

c) Prevention of acid conditions by providing reserves of alkalinity. The last two (b) & (c) can be described as wet stage regime and is applicable for a short period of 1 - 3 days and up to maximum 6 weeks max. under adequate concentration of chemicals. For 1 to 2 days shutdown, 10 ppm hydrazine residual & pH 10 with ammonia is competent for storage, and for an outage of 2 to 5 days hydrazine residual is needed to the amount of 50 ppm and for 5 days to 6 weeks, 300 ppm hydrazine residual with 10.5 pH with ammonia is essential. All these chemicals are needed to be drained before light up of the boiler again. However, during start up, maintain 2 ppm of hydrazine residual in the storage solution, provided the corrosion products are on acceptable limit.

3. **Checking of Mechanical Seal of Boiler Feed Pump:**

   After each cold start of boiler, it is necessary to check the mechanical seal of the boiler feed pump. Frequent outages of BFP are taking place because of failure of the mechanical seals due to fine rust particles, coming from pre- boiler system when subjected to corrosion mostly offload. Wet preservation of deaerator with 100 - 150 ppm of hydrazine or hot preservation with hot air should be done.

4. **Hydrostatic Test:**

   Whenever any hydrostatic test is to be done for a boiler & its parts, it is always advisable to drain the previous solution properly and flush with hydrazine/ ammonia dosed water to bring the remnant water free from salts & fill the boiler with 2 ppm N_2H_4 residual with pH 10 made with ammonia, after which only hydrostatic test has to be performed. Still it is necessary to waterfill the superheater (backfilling with DM water) beforehand to avoid contamination of superheater from any salt & corrosion products. After the test is performed, draining of boiler has to be done up to normal level and brought up in usual manner with the hydrazine/ammonia dosed water already available in the system. This test is usually done at the pressure 1.25 times of the rated one.

5. **Stator Water Cooling:**

   In many electricity generating units, the stator portion has provision to be cooled with either DM water or condensate, the quality of which also if not monitored may lead to generation loss. Corrosion products in such system can accumulate on unprotected mild steel components of stator cooling water system, especially in water boxes and in the conductor waterways leading to increased risk of electric flash over. Tube blockages due to corrosion products are separate factor. Among the corrosion products, iron oxides are now minimised due to improvement in design & replacement of mild steel in all vulnerable points with copper or stainless steel. Water boxes also have been replaced by electrically insulating hoses. Hence, attention is now being paid mostly of corrosion of copper in stator cooling circuit.

Stator cooling water is contained in a closed loop system that cools the copper stator bars in water cooled generators. The holes through which the water flows are narrow. Unimpeded flow through all stator bar openings is critical to operation of the generator. Overheating of stator bars can result in reduced generating capacity or even catastrophic failure of the generator. It can be seen from figure that there are two regions where the copper is best passivated with minimum corrosion. With low dissolved oxygen, there is zone where corrosion is inhibited by a passive film of cuprous oxide (CuO), and in another zone, there exits higher dissolved oxygen with passive film of cupric oxide (CuO). Although corrosion rates between these zones are moderate, it has been found that changes in oxygen level may cause sufficient condition from cuprous oxide passivation mode to a cupric oxide or vice versa for increased corrosion with detachment of oxide film. However, the low oxygen regime is simpler to control and is generally recommended.

Apart from pH, conductivity & hydrogen content are necessary to be monitored to avoid corrosion. If condensate having low DO is used in cooling circuit, controlled ammonia is also required to have non adverse effect on copper. If DM water is taken, the same is introduced into cooling circuit at a point preferably below expansion tank maintaining 250 - 300 mm Hg column vacuum with proper ejector working continuously. Further to maintain requisite quality of cooling water as recommended by manufacturer, a polishing unit is generally provided in bypass circuit to remove corrosion products & other dissolved salts from the circuit. The mixed bed ion exchange polishing unit is generally used preceded with 3 micron strainer. In any way, corrosion products have to be removed from the cooling system, to avoid tube blockages and electrical flash over.

6. **Condensate Polishing:**

The high electricity generating units preferably of 500 MW and above, where boiler feed water impurity can cause heavy loss of generation or where any sea water condenser cooling programme is there, it is always preferable to have condensate polishing to purify the condensate from corrosion products and some dissolved impurities like Cl, SiO_2, SO_2, Na, Fe, Cu, etc. so that feed water quality has no adverse effect on effective running of boiler & turbine. Polishing amount in different utilities vary from 50 - 100% of the condensate continuously and types of polishing units as adopted are of two types as follows:

Mixed Bed Ion Exchange:

The mixed bed is generally of deep bed alone or preceded with either cation exchanger unit (in case more ammonia or Na or other cationic impurities need to be removed), precoat, or magnetic filters, etc. (to remove corrosion products). The resins in deep beds are generally externally regenerated to avoid contamination by wrong regenerants to individuals due to poor separation of two resins (cation & anion resins).

The principle of operation of this mixed type of polisher is just like mixed bed in DM plant, but here care has to be taken for selection of resins to be of premier grade, more than 12% DVB crosslinked to withstand high pressure drop caused by flow velocity of 90-150 m/h approx. Apart from this, some particulate loadings on resin bed cannot be ignored together with resin working with high temperatures.

However, the mixed beds operating on H^+/OH^- cycle removes ammonia from the condensate together with other ions, adequate ammonia dosing arrangement on ex-condensate polisher is essential to provide suitable alkalinity to ex-polisher water although the requirement is gradually reduced due to resins in mixed bed being ammoniated gradually by the ammonia normally present in the influent condensate based on this concept, some utilities having all ferrous system in feed-steam cycle metallurgy and employing a pH of 9.4 to 9.6 are already in practice of operating the mixed bed units beyond ammonia breakthrough to extend the operating cycle and thus using ammoniated resin in the polisher to avoid or minimise further dosing of ammonia at ex-polisher. After exhaustion over a week or so, cation resins are additionally regenerated with ammonia also after the normal regeneration with acid. Use of ammonia (0.5%) is further made in rinse recycle of the regenerated anion resins through regenerated cation resins during its ammoniation. This operation aids in loosening the crud from the anion resin, and hence ammoniated resin has additional advantage of cleaning to give lower iron leakage from ammoniated mixed beds.

However, operating a mixed bed polisher on NH_4 /OH mode (though extend the operating cycle) is quite different from operating in H/OH mode due to:

1. The kinetics are different
2. The leakage are a function of pH & resin phase composition. To obtain sodium and chloride leakage of 1 ppb & 5 ppb, the following resin phase composition must be achieved

Thus, a very high degree of site conversion is required to obtain low ionic leakage from ammoniated mixed beds, which can be achieved only by: (a) employing high levels of regenerant (generally above 200 gm/l) and (b) eliminating cross contamination of resins. In H/OH cycle, it is important to prevent anion resin contamination of the cation resin, but in ammonia cycle, both anion contamination of cation resin and cation contamination of anion resin are of importance. Further, in presence trace quantities of organic materials as fulvic or humic acids are able to readily displace sulphates & chloride in a high pH environment. Alongwith, if sulphuric acid is in use for regeneration, this is either absorbed or held on the resin as HSO - in the anion portion gradually & deteriotes the quality of treated water adversely. With ageing of anion resin, retention is more. Iron acts as catalyst in oxidation & breakdown of normal sulphonate styrene dininyl benzene copolymer. This will cause fouling of anion exchange resins. However, high cross linked cation resin are more resistant. Thus the possibility of poor chloride removal from anion exchange resins always exists. Yet it is necessary to have a very high degree of cation and anion resins separation by suitable method & external regeneration to regenerate each of the resins in separate vessels. Additionally, it is important to always maintain a spare bed in the H+ form or have a bed initially operating on the hydrogen cycle with subsequent ammonia dosing system at ex-polisher to be ready for clean up in case of condenser leakage, otherwise high amounts of ammonia may develop in the polisher effluent together with chloride leakage. This means that for a unit having frequent shutdown and start ups, polishing unit operation could be both in hydrogen and ammonia cycle and during start up condition, it must be in hydrogen cycle.

Ammoniated mixed beds regeneration frequency is generally less than hydrogen cycle mixed beds. CO_2 loading on the condensate further lowers its run. Also, for operation through ammonia cycle, the sodium content of the cation resin must be reduced to less than 1% residual in order that a low sodium leakage can be achieved, whereas in H/OH mode, approx. 40% sodium can be present before sodium leakage exceeds 1 ppb.

The scheme flow & regeneration techniques of deep mixed bed are illustrated in following figures respectively.

In the scheme flow as in figure 7, it is indicated that resins are regenerated externally by transfering the resins from the operating vessel to a regenerating station in a vessel called 'separator & cation regenerator', where resins are separated by backwash & the top anion resins are further transferred to another vessel for regeneration, called 'anion regenerator & resin storage vessel'.

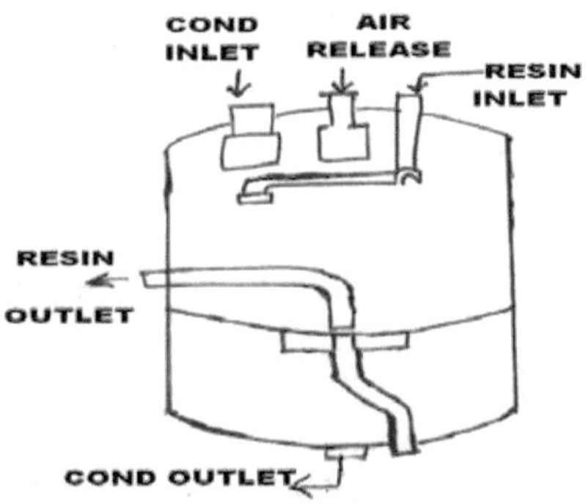

DEEP BED CONDENSATE POLISHER

FIGURE 6

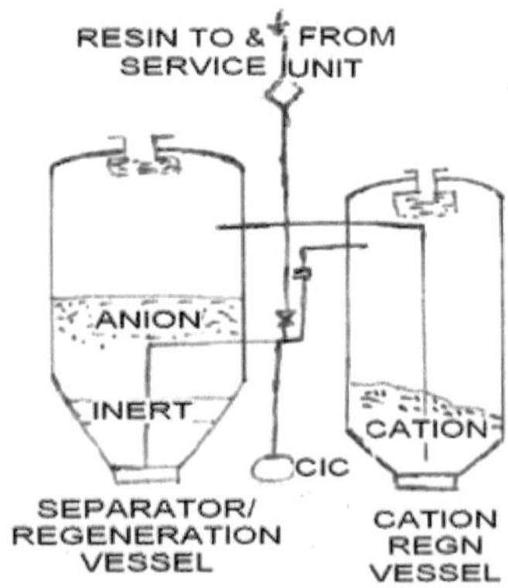

FIGURE 7

The transfer anion resins are generally in two steps-transfers of anion resins 150 mm above interface & then high backwash followed by low backwash of bulk cation resins to separate further anion resins and thus transfer the whole anion resin. Regenerated cation is put first in part for mixing in anion regenerator cum resin storage vessel, and after due mixing is sent back to the mixed bed operator. Another resin separation & regeneration technique is illustrated in figure 7, in which help of inert resin is taken for effective separation & the bottom of the regenerating vessel are made of cone type. The cone arrests some mixed resins, which are either thrown away or transferred to store in a separate vessel, which when after few regeneration become substantive, then these are again taken into separator to separate cation & anion resins individually.

The condensate circuit for 500 MW is generally at 46 bar & regeneration system at 10 bar. High pressure in transfer lines raises an alarm and stops the sequence. Hence, before transfer, it needs to be depressurised & resin bed fluidised to release bed compaction. Then resins in the form of slurry are mostly transferred at 12 m^3/h & could be up to 0.5 km distance.

Ammoniation of the resins could be both on & off line and are done as indicated in figure 9. Previously ammoniation was performed by applying'once through'with 0.25 to 0.50% ammonia for a period of 3 to 4 hours, but now "recycle ammoniation" is adopted to save ammonia. Use of ammonia in the anion rinse operation aids in loosening of crud from the anion resin and helps guarantee lower iron leakage from ammoniated mixed bed. Also, in view of operation of mixed bed on ammonia cycle taking longer time before next regeneration, it is likely to have more crud loading also, which must be removed during regeneration. Hence, the mix is given either resin air scour & rinse (Rasar) or special patented scrubbing procedure called 'ABRO'-(Air bump rinse operation, which is a cyclic operation of air for one minute followed immediately with flow rinse for two minutes, and this is repeated depending on crud loading. After that the normal backwash & separation of resin follow.

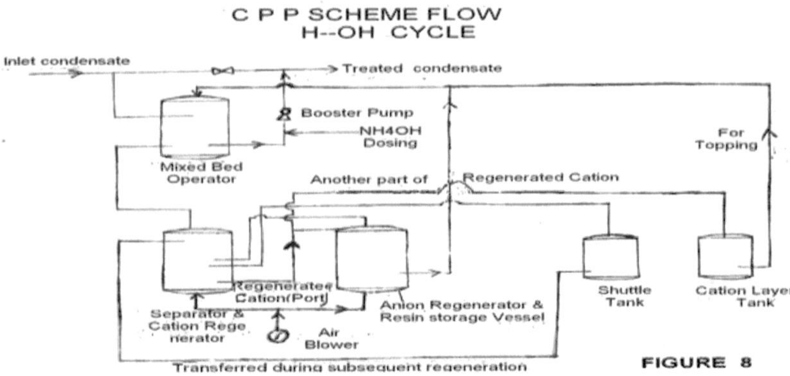

FIGURE 8

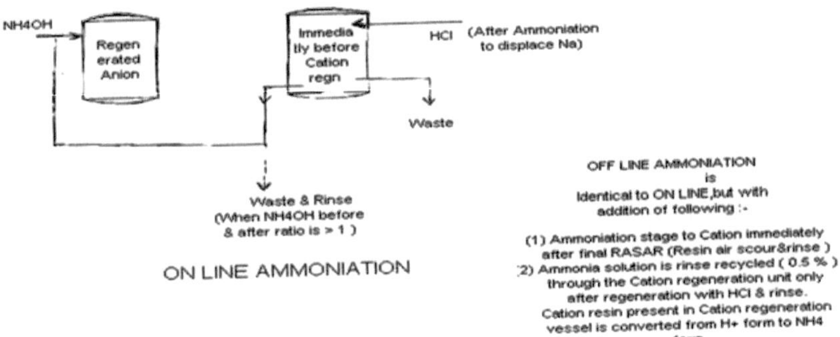

FIGURE 9

B. Pre-coated Filter:

Ion exchange resins have further been used in condensate polishing by giving a coat of expandable powder or micro bead resins, which act not only as a very fine filtering medium to trap particulate matter, but also removes soluble contaminants by ion exchange. When the pre-coat media is either clogged or exhausted, it is simply disposed off, and the cartridges are coated with fresh slurry of powdered resin. However, in operation of this powdered precoat, it is necessary to keep the vessel always on pressurisation; otherwise there are chances of pre-coat material falling. This is necessary even when the generating unit is on or offload. Again one has to be very careful in applying precoat uniformaly. Uneven coating may hamper the performance. The functional, coating and cleaning operations of a pre-coat filter are shown at figures:

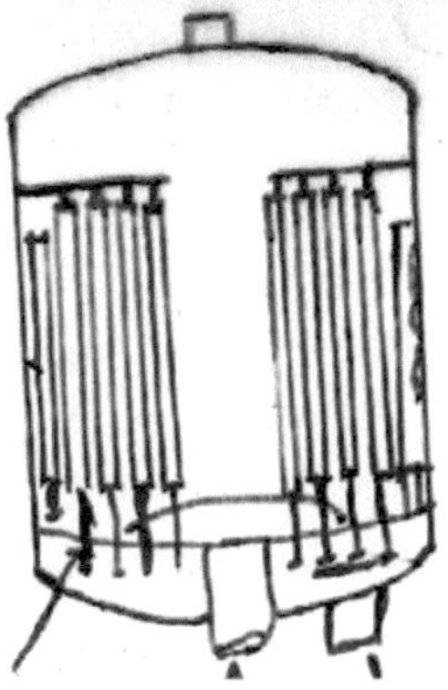

FIGURE 10

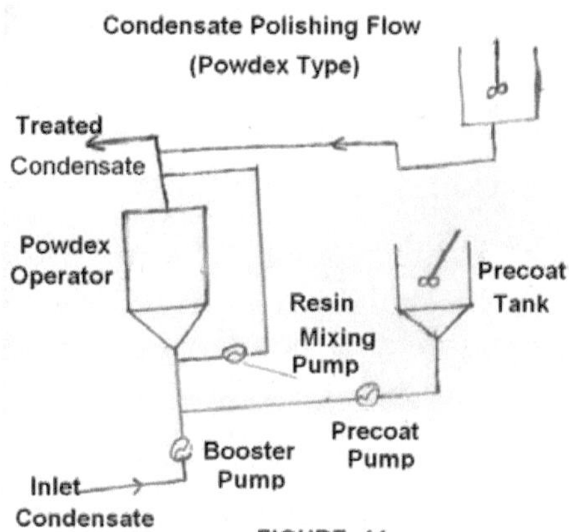

FIGURE 11

7. **Steam Washing of Turbine:**

This is required when too high deposit on turbine blades & its accessories are suspected to effect the load (in some installations indicated by high curtis wheel pressure also). In 210 MW, steam washing starts when main steam temperature reaches 225 ^{0}C with steam pressure 27 - 30 bar & reheat steam of 160 ^{0}C. Effectiveness of steam washing is determined by extent of presence of salts & dissolved (DO), in the main condensate, which is drained out:

SiO_2	less than 40 ppb
Fe	less than 20 ppb
Cu	less than 5 ppb
Hardness salts	less than 1 ppb
Dissolved O_2	less than 10 ppb
Oil & Grease	less than 0.3 mg/kg

Normally completion of steam washing operation of HP cylinder & IP cylinder takes 2.5 to 4 hours respectively. During steam washing, the condensate should be drained totally (arrangement must be like that) and extra DM water is fed to deaerator.

While onload reduction, one has to be careful also to note whether there had been blade washing too and hence lab must examine it before boxing the unit. Continuous occurrence of this together with loss in generation will actually determine the need of steam washing of turbine, which however should not be too frequent.

8. **Checking vibration of turbine rotor:**

   Osion Cinorbr earing of turbine may cause sliding of axis and vibraton may occur. Moisture and corrosion products in turbine lubricating oil are necessary to be kept under good acceptable limits. Accordingly, vibration of turbine rotor needs to be checked.

9. **Chemical Cleaning and Passivation:**

   While operating a boiler, when a change in operational chemistry including occurrence of hideout, persistent condenser leakage, debris found in boiler tubes & drum, onloadcorrosion with extensive retubing and observation of increased pressure drop in once through boiler indicating deposits in tubes, etc. are noted it is necessary that boiler & its parts including economiser are cleaned suitably with suitable chemicals predetermined as per the nature of deposits found. Superheaters, reheaters, condensate & feed systems are not included in the circuit for cleaning. Even when no significant operational abnormal chemistry are noted, there should be routine cleaning of the plant either during each overhaul with modification if any, or on finding the magnetite thickness/deposits thickness above 50 micron. The greater thickness generally gives increased risk of corrosion. As a general rule, 25 micron thickness in 25,000 hours operation is taken as satisfactory for taking up cleaning operation. In other words, chemical cleaning is used to control oxide & deposits thickness & thereby to limit the concentration factor. Chemical cleaning of a boiler is generally found necessary when oxide thickness growth is found greater than 50 micro metre or, internal water walls deposits are found between 15 - 35 g/ft^2.

   Similarly during precommissioning of boiler & its parts, if the internal surfaces of steam water circuits are not found free of foreign matter including mill scale, grease, oil & dirt, the chemical cleaning is required to commission & operate the plant with good protective magnetic film. However, if all parts are cleaned at the factory (as now generally being done) with steps to seal them & preserve before sending them to site and then are stored properly at site also to finally erect them as per standards minimising the ingress of foreign matter into the system, then after erection, cleaning can be limited to flushing with provision to remove temporary protective coating & preservatives, if any. In view of nowadays, boiler manufacturer usually using cold rolled steel, possibility of presence of mill scale is also ruled out.

For removal of oil, dirt, internal protective coating, preservatives, etc. cleaning is done by 'alkali boil out' in which tri sodium phosphate, NaOH, detergents, etc. are used, but for removal of metal oxides as usually found, hydrochloric acid, citric acid, hydrofluoric acid, ammoniated citric acids, etc. are used and is called 'acid cleaning' process. In view of hydrochloric acid causing damage to tubes particularly in the difficultly drainable portion or under deposits, cleaning by this acid is now a last choice with much more attentive procedural steps. Handling of HF acid is also very difficult at site, but can give a good cleaning effect including of those portions having alloy steel like superheaters, reheaters, etc.

The recommended methods of chemical cleaning step by step are as below:

Flushing

The system mostly individually is flushed vigorously with filter water/DM water at maximum possible velocity to remove foreign matter.

Caustic Soda Cleaning:

This is applicable to such system like low pressure (LP) Feed heaters and others, including low pressure boiler drums, etc. which have been cleaned & coated with temporary protection with Ceralac or Trilac. The alkali solution of required strength (0.5% for ceralac & 0.25% for Trilac) is circulated at 40 to 60 ^{0}C for 4 hours and then flushed till pH 8.0 initially with filter water & then DM water and finally preserved with 50 ppm N_2H_4 with no ammonia where feed system has copper metallurgy also, otherwise supplemented with ammonia to have pH greater than 9.5.

Alkali Boil Out:

This is applicable to boiler & economiser to remove greasy & siliceous matter and at its post commissioning removed dirt after annual overhaul and to repassivate with hydrazine before recommisioning.

The boiler is filled to working level along with economiser with 1000 ppm trisodium phosphate solution and approved detergent/surface active agent/non-ionic surfactants. By firing, the boiler, drum pressure is raised gradually to 40 kg/cm^2 within 8 - 10 hours, during which intermittent blowdown for 30 seconds every hour are given. The final pressure of 40 kg/cm^2 is maintained for 24 hours with specified concentration. If the silica & oil content in the blowdown water is still found on increased side, a second boil out with filling with fresh chemical has to be repeated in the same way till silica & oil content are either less or constant. The boiler is then drained while still hot & then cooled and opened for inspection to remove loose debris & arrangement for onward acid cleaning. Apart from removing oil, grease, silica, etc. trisodium phosphate forms some complex with iron, aluminium, zinc, calcium & magnesium compounds, which are removed from system through blowdown. Surface active agents like Lissapol activate the cleaning efficiently.

Citric Acid Cleaning:

This is applicable to LP & HP feed water pipe works, LP feed heaters, superheaters & reheaters during pre-commisioning and condensate & feed water system excluding economiser in post commissioning In boiler & economiser also this process is required generally after HCl/HF acid cleaning except under post commissioning when deposits having considerable amont of copper deposits too need to be removed. The cleaning process is undertaken by circulating a suitably inhibited 3% citric acid (solution having less than 2 ppm chloride) and sufficient ammonia to give pH 3.5 - 4.0. In absence of alloy steel material in circuit, 5% ammonium bifluoride is also added. The circulation temperature is maintained at 75 ^{0}C and the period of circulation is until the iron content is found constant, the concentration of citric acid should not fall below 0.5%. After draining the system, initial rinse is done with 0.2% citric acid ammoniated to pH 3.5 - 4.0 & then with DM water till pH 6.0.

However, citric acid cleaning is not effective in presence of thick layer of scale (greater or equal to 300 gm/m^2 Fe_2O_3). The solution on saturation with iron compounds is broken down with precipitation of iron citrate or hydroxides on the cleaned surface of metal, followed with sharp drop in temperature of solution (10 -15 ^{0}C) or by a reduction in the circulation rate by choking of tubes. Cleaning is therefore recommended in two stages - first with 2.5 - 3% and then with 1 - 1.5% solution of ammonium mono citrate.

HCL Cleaning:

This is applicable to boiler & economiser and is always followed after alkali boil out during pre-commissioning or as per requirement during post commissioning. This is being done by circulating 5% HCl @ 1 m/sec, mixed with 0.5% ammonium bifluoride (to activate and to take care of some siliceous matter if present) and approved inhibitor (like rhodine special, etc. of required strength), at 60 ^{0}C for 6 - 8hours or till iron content is constant, but before that it is ensured that superheater & reheater are suitably primed with ammoniated DM water. The boiler & economiser is then drained with back flushing of superheater and rinsed repeatedly with a solution of 0.2% citric acid ammoniated to give pH 3.5 - 4.0 until chloride content is below 2 ppm. It is again ensured that superheater & reheater are thoroughly primed and then boiler through economiser is filled with 3% citric acid ammoniated to pH 3.5 - 4.0 with approved inhibitor of suitable strength, and the solution is circulated at not less than 90° C till iron content is constant. In no case strength of citric acid during last stage of circulation should be less than 0.5%. The plant after draining is again filled & flushed with 0.2% citric acid ammoniated to pH 3.5 - 4.0 and finally with DM water till pH 6.0. Finally, the system has to be filled immediately with 300 ppm N_2H_4 & 50 ppm NH_3 and kept under this condition for minimum 24 hours.

However cleaning with solution of HCl has following disadvantages:

1 It tends to diffuse under the deposit and to peel it off in shells and thus large quantity of suspended matter may be present in the cleaning solution, which can cause clogging of boiler tubes partially at bend and low flow area. At this portion, even after washing there are chances of HCl and chlorine fumes to remain and cause damage to tubes.

2 It cannot be used on plant containing nitrided components, which are immediately corroded by HCl

3 It cannot be used on plant containing components made of austenitic/alloy steel, thus parts of plant remain uncleaned, which are later cleaned by steam blowing, which is time consuming.

Hydrofluoric (HF) Acid Cleaning:

Hydrofluoric acid attacks the deposits from the surface in which the particles are dissolved & not dismantled (as in case of HCl). Thus, if the acid is suitably inhibited, there is no selective corrosion on most of the materials used in a water-steam cycle. Hence, boiler including superheater can be cleaned at one go, saving lot of time in bringing up the generating unit. Solubility of iron oxides in 1% HF solution is several times higher than HCl solution of 5%.

HF and dissolved iron in waste water can be neutralised with calcium hydroxide as insoluble precipitate and thus waste water disposal is not problematic. If plenty of water having considerable M-alkalinity is available, the same also can be used to neutralise the waste HF acid. Hence, HF cannot be used when the deposits are reported to contain calcium ion in appreciable amount. Insoluble CaF formed thus is difficult to remove from tube surface, which can disturb protective layer formation later on. Also use of HF alone is restricted at high iron (i) oxides content in the deposits, in which case use of reducing agent like Tin (ii) chloride/ascorbic acid or even formic acid can be used which binds Fe (iii) ions in a complex form. Additionally, HF can open up leaks to welding inclusions and there is no test data to establish that fluoride is safe for austinitic steel.

Inhibited 1% HF is injected down stream of boiler feed pump & dosed to maintain 1% strength at the last outlet point. Drum internals need not be removed and superheater/reheater can be taken into circuit together. Velocity of injection is kept 0.2 m/sec across the largest heating surfaces, which in most cases are reheaters. After 1 hr of injection, there is rise in iron concentration reaching its maximum after 1.5 hours to decay again after 3 hours. HF concentration falls to a value of 0.7% after 2 hours i.e. at a time when iron concentration is decreasing, and acid injection is then discontinued. After 20 minutes, later when 1% HF is obtained in the effluent, it marks the cleaning process complete. The acid clean solution is then displaced at a high velocity to waste. Thus, by this once through cleaning, the effluent waste is either collected in neutralising tank or neutralised with $Ca(OH)_2$. Neutralised waste is insoluble CaF_2 safe to be discharged. If cleaning is done by circulation method, 0.5% HF solution is used and circulated till iron values are constant at sufficient acidity of the medium. After completion, the rinsing is done with DM water till acid free and then immediately with hydrazine/ammonia treatment as in case of HCl cleaning above.

Particular care must however be taken when using high% of HF. In diluted form below 2%, it does not however seriously attack human skin and neither steel or concrete.

Finally, whatever be the choice of acid to clean boiler tubes, care should be taken that the same does not take away more than 0.4 mm material per acid clean. Usual boiler tube material is 7 mm thick.

Cleaning Deposits Having Copper:

Copper does not dissolve either in HCl or HF. Rather it precipitates from cleaning agent in form of metal on cleaned steel surface. Hence, immediately after termination of acid cleaning, washing with 0.7 to 1% citric acid is followed till iron content is constant (usually achieved in 30 - 40 minutes). The citric acid is then neutralised with ammonia to about pH 10. Then 0.1% oxidising agent ($NaNO_2$, $KBrO_3$, $KClO_4$, $Na_2Cr_2O_3$, ammonium peroxide sulphate, H_2O_2, etc.) is added to deactivate bright metallic surface. Increasingly however the use of H_2O is common, which brings about additionally a passivation of the clean steel surfaces.

The presence of copper in the deposit is also decisive for the selection of inhibitors. Amines based inhibitor commonly used for steel are emulsified by copper & become ineffective. In place thiourea derivative should be selected.

However, if copper is present in the deposit in higher amount (more than 5% or so), it is always preferable to remove copper first with properly inhibited citric acid before going for HCl/HF acid cleaning. Usually two stage citric acid treatment is recommended and during the second stage, conditions are set up for chelation of copper compounds. After first stage circulation of citric acid for 1 - 1.5 hour, additional concentration of 0.5% is made and then pH of cleaning solution is increased to 10 with ammonia with a subsequent reduction in temperature to 60 - 65 ^{0}C and the introduction of 0.5 - 1% of an oxidising agent (usually sodium nitrite).

Passivation of Cleaned Surface:

To prevent subsequent corrosion of the acid cleaned surface, it is necessary to create protective layer conditions of magnetite (0.01 to 4 micro metres) soon, so that the plant can be put to safe operation even if there is delay in putting the plant in operation due to some local problems. The passivation is done in two stages after acid cleaning & rinsing is complete, the boiler & economiser are filled with 200 ppm N_2H_4 + 50 ppm $NH_{3.}$ The boiler is then fired to raise pressure to 40 kg/cm^2 and maintain for 24 hours, with make-up water also having 50 ppm N_2H_4 & 50 ppm NH_3. The boiler is then blown empty at highest possible pressure, but not at less than 3 kg/cm^2. It is capped with nitrogen & allowed natural cooling. The drum is then opened for inspection of loose debris & for their removal. Drum internals and instrumentations are then restored, superheater & reheater plugs, if any, are removed & boxed up. This is called first stage passivation. The boiler is then filled again with 200 ppm N_2H_4 with pH 9.6 made with ammonia. It is lighted up again supported with make-up water having 20 ppm N_2H_4 & NH_3 (for pH 9.5). Drum pressure of 40 kg/cm^2 is again brought after firing and maintained for 24 hours and after that drained hot with nitrogen capping and then allowed to cool naturally to ambient temperature. This marks the 2nd stage passivation and then the boiler is preserved dry. If steam blowing to clean superheater & reheater is not carried out within a month then second stage passivation needs to be repeated. After annual overhaul of boiler also, this 2nd stage passivation needs to be repeated before loading the boiler for operation. For a conventional boiler tube of about 7 mm wall thickness, and an advised design corrosion allowance of 5% for chemical cleaning, the tolerable number of post service cleans would be 10 with citric acid & 7 for HCl. So there is ample margin for maintenance cleaning at 3 to 4 years intervals in the 30 year design life of the boiler.

Means of Cleaning Condensers & Coolers:

The selection of appropriate solvent and inhibitor varies from case to case, as it is considered on the following factors:

a) Qualitative & quantitative estimation of the deposits to be dislodged
b) Material of construction of heat transfer elements
c) Flow path characteristics of the equipment

Possibility of sulphate precipitation in case of excess acid is used and results in calcium scale due to decomposition of sulphamic acid still remain at high temperature, greatly slowing the cleaning rate. NaCl: Sulphamic acid (1:2) enhances the cleaning rate, but this defeats one of the major benefits of sulphamic acid, namely that it is a low chloride material. Personal exposure with formic acid needs to be avoided; permissible limit in air is 5 ppm with immediate dangerous level 100 ppm.

Because of limiting solubility of calcium formate, the cleaning must be designed for 2 or more stages of acid addition, if calcium based scales indicate more than 10% formic acid requirement as per stoichiometric calculation.

A reducing agent is generally required to assist in dissolving corrosion products. Ammonium bifluoride is frequently added to provide effective performance in removing silica, calcium & magnesium silicate & silt.

After the chemical cleaning operation of boiler and economisers are finished including with passivation/preservation programme, steam blowing operation normally follows particularly in precommissioning stage of boiler, to clean the main steam piping, superheater, etc. so that none of the impurities like particulate matter remain in the system, to reach later to the turbine during operation. When reheaters are present, this step is usually done in two stages - first the main steam piping and superheater are blown and then, the main steam piping, superheater and reheater are blown. The most effective cleaning is accomplished when the blowing steam force exceeds maximum operating steam force. A 1.6:1 force ratio has been found to provide satisfactory cleaning of particles from the piping. Continuous blowing is used, but intent blow periods of several minutes each are more common.

In order to assess blowing effectiveness, blow steam may be directed onto a target plate installed in the main steam line. Successive blows are made until solid particles erosion is no longer visible on the target plate, which some times take more than 24 hours. During steam blowing operation the boiler must be treated only with all volatile chemicals, viz., ammonia and hydrazine

Sampling Points: Concentration of 0.5% w/v needs to be maintained.

Sampling points for all vital points need to be provided (as per fig. 1) to monitor the quality of each section of feed water-steam cycle and should also be representative of the system for which the same are provided. Alongwith, all the sampling points should run continuously also at constant optimum flow. The maximum length of a sampling line should not exceed 25 metres. If the length is more, iron contents obtained in the sample could be more by 80 - 90% of actual values. Regulating of sample flow must be adjusted by the value after the sample cooler. The material of construction for sample collection should be the same as that of system & the velocity in the sample pipe must be maintained in the ratio of 0.25 - 0.5 w.r.to that in the system, so that the sample is able to reach within 3 - 5 minutes to sampling panel @ 1.5 -1.8 m/s. In case bend is necessary in the sampling line, the pipe after the bend must not be horizontal or upward, but should slope downward slightly.

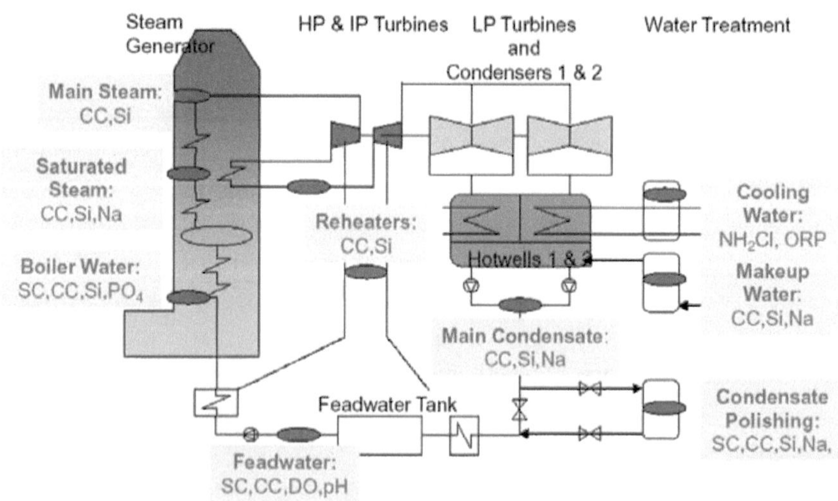

Plant Monitoring

Boiler Feed Water

Feed Water Treatment Type	Process
AVT (all volatile treatment) Feedwater	Feedwater pH is maintained in between 9 to 10 with addition of ammonia. Dissolved oxygen has to be removed by adding oxygen scavenger (hydrazine) or with the use of degasifier.
CWT (combined water treatment)	Feedwater pH is maintained in between 8 - 9 in addition of ammonia. Dissolved oxygen content between 150 to 300 ppb.
OT (oxygenated treatment)	OT (oxygenated treatment) – Feedwater is not conditioned to raise the pH. The pH value is kept in between 7 - 8. Controlled injection of oxygen maintains a level of 150 to 300 ppb.
CC (cation conductivity)	For identifying low level contamination by potentially corrosive anion contaminants such as chlorides, sulphates and organic acids.
pH	Monitor and controlling the alkalisation process control function.
Dissolved Oxygen	1. Controlled injection of oxygen for OT. 2. Monitor the removal of dissolved O_2 for AVT. 3. To determine dissolved oxygen removal in the deaerator.
Oxygen Scavenger	Monitor and control injection of oxygen scavenger.

Boiler Water

Treatment Type	Process
EPT (equilibrium phosphate treatment)	Phosphate residual is maintained at 0.2 to 2.5 ppm. pH range at 9.0 to 9.7
Caustic Treatment	Addition of sodium hydroxide to the boiler water. Boiler water pH controlled at 9.4 to 9.6. Maintain NaOH level at 1.0 to 1.5 ppm.

Boiler Water

Treatment Type	Process
pH	Low pH in the water attacks the steel, thus the pH conditioning process has to be closely monitored.
SC (specific conductivity)	Controlling and monitoring of the chemical treatment.
Silica	To monitor silica contaminants level in boiler water.
Phosphate	To monitor the residual phosphate in boiler water.

Main Steam

Parametres	Process
CC (cation conductivity)	For identifying low level contamination by potentially corrosive anion contaminants such as chlorides, sulphates and organic acids.
DC (degassed conductivity)	Measurement of CO_2 contamination 1. To determine air or cooling water leakage. 2. To reduce the startup time especially for cold start up by achieving the conductivity level determined by turbine manufacturer.
Silica	To monitor silica carryover in saturated steam to protect turbine blades.

Parametres	Process
Sodium	To detect solids (sodium salts) present in steam. Sodium deposits on turbine blades will cause vibration and lead to damage.

Condensate Polishing

Treatment Type	Process
Deep-bed Condensate Polishing	Hydrogen-hydroxide form, mixed bed polishing.
Powdered-resin Condensate Polishing	Cation-anion resin, mixed bed polishing.

Condensate Polishing

Parametres	Process
CC (cation conductivity)	For identifying low level contamination by potentially corrosive anion contaminants such as chlorides, sulphates and organic acids.
Silica	To detect ion exchanger exhaustion at polisher.
Sodium	1. To detect sodium ion (impurities) breakthrough and ensure product water is pure. 2. Used for condenser leakage detection.

Feed Water - Steam Cycle Failures

There have been nowadays improved metallurgy & design of feed water-steam cycle components along with practice of adopting ultra pure quality water feed as make-up in almost all TPS having particulary high pressure boilers, yet problems of failure cases persist. The problems, however, are generally limited to erosion, corrosion, heat imbalances, operational practices and overall transient chemistry due to frequent start ups & shutdowns. Bad cycle design has also been found responsible for corrosion & excessive scale/deposits problems as below:

Bad Cycle Design:

1. If sufficient purity removal provisions (viz., blowdown, condensate polishing, start up recirculation, deaeration, etc.)
2. Low make-up system capacity
3. Use of mixed metallurgy (Cu alloy in feed heaters)

Bad Component Design:

1. Boiler with poor circulation
2. High heat flux
3. Piping & feed water heaters with high flow velocity 4urbuleTnce causing flow accelerated corrosion
4. Under sized deaerators unable to reduce non-condensible gases
5. Poorly deaerating condensers, turbine discs & blades operating with high stresses or near resonance condition leading to stress corrosion & corrosion fatigue
6. Inadequate boiler drum internals

To prevent all these, certain operating guidelines including quality maintenance of the cycle chemistry have already been discussed on previous pages. Yet erosion & corrosion problems as being faced are necessary to understand in detail, which are discussed as follows:

Erosion

High velocity water & especially water/steam mixture cause erosion in feed water systems. Sometimes problems of erosion also occur at moderate average flow velocities when a sequence of bends causes a significant increase in local velocity. Equipments vulnerable to erosion includes turbine blades, low pressure steam piping, heat exchangers subjected to wet steam and also feed water & condensate piping subjected to high velocity. Damage normally occurs where changes direction flow. Oxygen laden water/steam mixtures further aggravates and creates familiar thinning pattern including pitting at economiser inlet. CO_2 laden water/steam mixture results into end point erosion-corrosion of turbine blades of LP region last row. Gaseous laden (oxygen & ammonia) water/ steam mixture again have erosion or corrosion at deaerator venting region.

The most commonly encountered erosion problems occur at hairpin bends in steaming economiser, thinning the elbow leaving a characteristic reverse horse shoe imprint.

Futher, erosion of high pressure end of turbine is usually caused by solid particles like iron oxides and erosion of IP & LP turbine blades by water in steam, formation of which can be witnessed earlier, if operated below design inlet steam temperature or at low load due to condensation in these stages. Direct impingement of wet steam on to condenser tubes are also causing thinning of tubes by erosion. Visually the affected surface is brighter & cleaner than the normal ones, and is generally seen on top bank of the condenser tubes where turbine exhaust steam makes contact with the tube bundles. A similar effect damaging baffle plates, etc. can also be seen, where heater drains are drawn into the condenser. In order to mitigate erosion problems, it is necessary to have good passive metallic oxide layer at these regions through a good water chemistry control, and if the problem is not resolved in this way, design & operational changes are required.

Over Heating

Similarly, failure due to over heating, if not found due to deposits (characterised by plastic flow of material), can be attributed either due to metallurgical failure or design defects. For a boiler tube, when conditions of deposits or blockages cause a rapid elevation in metal temperature to the range 9000^{0}C, plastic flow conditions are reached and a violent rupture occurs, characterised by thin sharp edges at open ends. Thin lipped bursts also occur in superheaters tubes when steam flow is insufficient due to a rapid firing rate during boiler start up. A long term scaling condition leading to a tube leak is usually indicated by a wrinkled, bulged/blistered external surface & a final thick lipped fissure or opening.

Cracks development due to overheating at and near the welds as commonly found in deaerator, have also been noticed in some installation, the only remedy remains is suitable steps of stress relieving of welds and minimisation of thermal & mechanical stress during operation.

Ripple magnetite formation on water side surfaces has been directly related to elevated temperature in the high heat flux areas & to boiler pressure drop loss problems. Above 300 ^{0}C though the general corrosion rate of SS is less than carbon steel, but caustic cracking will occur in ss, if a liquid phase is present. Ferrite steel tubes (11/4 Cr-1/2 Mo) in colder section of superheater did not fail, whereas AISI 304 HSS failed when steam drum contains 20 ppm Na (34.8 ppm NaOH) at 3000^{0}C called caustic stress cracking, characterised by blue oxide colouration on cracked sample. Ferritic steel tubes (1Cr-1/2 Mo, viz., AISI 304 HSS has limited resistance to creep rupture strength above 5750^{0}C, While austenitic steel, viz., 304 H have significantly better creep rupture strength. This is the reason for adoption of SS above 565^{0}C and below this 11/4Cr-!1/2 Mo steel are used which have sufficient creep rupture resistance.

Change in micro structure of metal (as generally noted in case of turbine blades) is generally due to temperature excursions during operation. Metal fatigue is also caused by changes in thermal cycle & thermal shocks including trip outs followed by cold, warm or hot starts of the units.

Corrosion

Corrosion is defined as the destruction of a material by reaction with its environment. There are broadly two types of corrosion, one being uniform and the other localised under which varieties of corrosion are found as per working conditions of environment. While uniform corrosion rate of the material, in which, metal is corroded uniformly over the entire metal surface, can be easily measured to predict accurately how long the metal will last in service, the damage caused by localised corrosion is quite difficult to predict.

Permissible rate of uniform corrosion at normal working conditions of boiler corresponds to 0.06 to 0.08 mm/yr, the rate of excessive corrosion being 0.08 to 0.20 mm/yr and emergency corrosion 0.2 to 0.4 mm/yr. For water economiser, rate of uniform corrosion at a wall thickness of 3.5 mm, may be adopted not more than 0.07 mm/yr. Maximum permissible depth of point corrosion is obtained by measuring the wall thickness on the damaged portion of the pipe remaining after corrosion, which is arrived at, by the formula

P	=w/k x 8.76, where P		= Penetrability in mm per year
		w	= Loss of metal in gm/M^2 hour
		k	= Coefficient of specific weight of metal in gm/cc (for steel 7.8 gm/cc)

The causes of corrosion in different regions of feed water-steam cycle have already been elaborated under pre-boiler, Boiler and turbine steam chemistry heads respectively, but it is still necessary to be aware of the different types of corrosion, TPS (thermal power station) are likely to face in their actual operation. Some of these types are discussed as follows:

Uniform Corrosion:

It is general attack on metal by water, acid/alkali, high temperatures, oxidation of metals, etc. due to which the metal becomes uniformly thin.

Galvanic Corrosion:

It is attack on anodic metal in contact with noble metal, viz., carbon steel with stainless steel. More active metal preferentially corrodes protecting the less active metal. Conductivity & composition of water & the areas of two metals determine extent of corrosion. Higher the conductivity of water, wider is the area of attack. Hence this type of corrosion is mostly faced more on cooling water side of TPS functions, which will be discussed separately along with other problems in these areas, under the chapter 'Cooling Water Chemistry'. Now limiting to corrosion here to boiler feed-steam cycle, anything that results in a difference in electrical potential at discrete surface locations can be treated to cause a galvanic reaction, as per the reaction shown in figure 12 (a) (keeping in view that at high temperatures, polarity reversion also may take place) due to:

a. Contact of dissimilar metals
b. Differential stresses in metal i. e. at welds in heat affected zones or use of different alloy in the welds.
c. Conducive deposits, viz., of Cu
d. Difference in temperature

HCl acid cleaning thus can cause galvanic corrosion unless Cu is prevented from plating on steel surface. Cuprous chloride ($CuCl_2$) is immediately deposited as metallic copper on steel. HCl also dissolves magnetite giving both Fe^{2+} and Fe^{3+}. $FeCl_3$ is very corrosive to both steel and Cu.

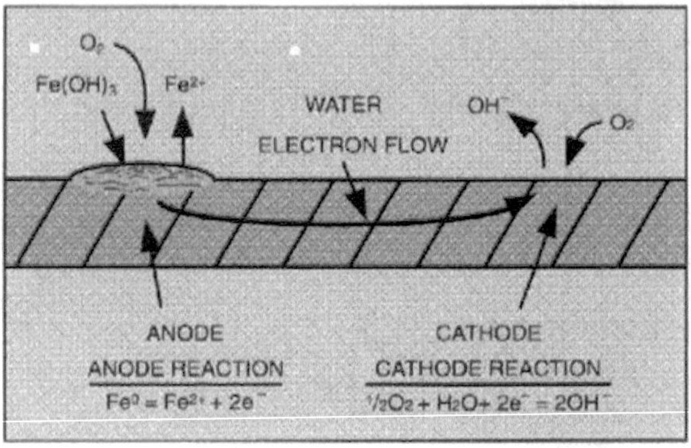

Simplified corrosion cell for iron in water.

Figure 12(a) Simplified corrosion cell for from in water

Caustic Corrosion (Gouging):

As shown in Figure 12 (b), caustic attack creates irregular patterns with few pits formation or metal loss. Depositions may or may not be found in affected areas, but is evident along with white salt deposits (Na_2CO_3) which usually outlines the edges of the original deposition area. Thinning of tube metal caused by this caustic attack creates irregular patterns — often referred to 'gouging'. Cause of caustic corrosion is mainly due to solid alkalis/concentration of NaOH at particular place e.g. due to steam blanketing, porous metal oxide deposits. Usually this type of corrosion is seen in economiser, where steam generation takes place beneath a deposit and free NaOH is present in feed water. Boiler with caustic attack/gouging often show excessive accumulation of magnetic oxides in low flow areas of drum & headers.

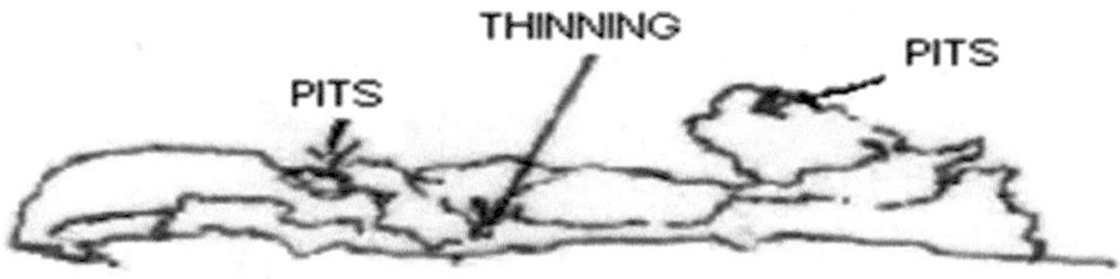

CAUSTIC ATTACK

FIGURE 12 (b)

Caustic Embrittlement:

This is a specific form of stress corrosion leading to 'caustic stress corrosion cracking' or 'inter crystalline cracking', for which the following three conditions must exist:

a) Metal must have high level of stress.
b) Mechanism for concentration of fluid exists, such as leakage.
c) Fluid must have embrittlement producing characteristics i.e. presence of chemical like NaOH.

Since, chemical attack is normally undetectable, failure occurs suddenly often with catastrophic results. $NaNO_3$ however in proper ratio with NaOH can inhibit imbrittlement, but in lower pressure boilers. Yet the proportions are different at different pressures.

Caustic cracking in boiler tubes can be seen as circumferential cracking as shown in figure 12 (c), while in other components cracks follow the lines of greatest stress. The crack progresses along defined paths or grain boundaries in crystal structure of metal as shown in the adjacent figure.

The cracks do not penetrate through crystal, but travel between them. Boiler tubes usually fail from caustic embrittlement at points where tubes are rolled into sheets, drums or headers. Embrittlement is possible on welded design also, viz., cracked welds, tube end leakage, etc. can provide mechanism by which drum metal may be adversely affected. Characteristics blue oxide colouration is associated with caustic cracking on SS.

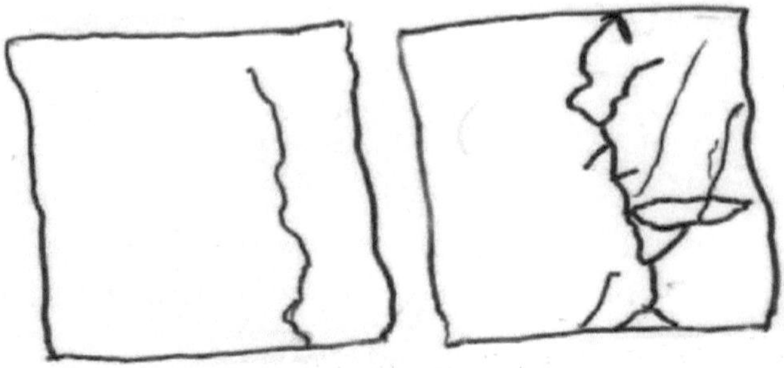

Crack starts from circumference   Crack travelling under grain boundaries

FIGURE 12 (c)

Stress Corrosion Cracking (General):

Stress corrosion cracking is the brittle failure of a metal by cracking under tensile stress in a corrosive environment. Failures tend to be transgranular, although intergranular failures have also been noted. Tube failures occur near stressed areas such as welds, supports or cold worked areas.

In other words, this type of cracking occurs from combined action of corrosion and stress, apart from the fact that the alloy is itself susceptible (e.g. type 316 SS in presence chloride). The corrosion starts right from improper chemical cleaning and are followed by the factors: (a) high dissolved oxygen levels (b) pH excursions in boiler water (c) presence of free NaOH (d) high levels of Cl and /or S.

And, the stress is due to static/dynamic effect of tensile strength synergistically, mainly caused by: (a) residual stress in the metal (b) thermal excursion (c) rapid start up & shutdown of boiler.

Corrosion fatigue & stress corrosion cracking of turbine blades and discs are commonly associated with S, Cl, and NaOH. Problems are more common in low pressure part of turbine, characterised by high stress, crevices and operating temperatures conducive to the condensation of concentrated solution of steam contaminants – with more impure steam problem, this can occur in high pressure part of the turbine also. While caustic cracking gives blue colouration on the cracked surface, sulphur compounds leave reddish brown surface. The problem of stress corrosion cracking can be mitigated by designs that prevent crevices, lower stresses and/or employ lower strength materials.

Crevice Corrosion / Attack:

It is attack at crevices, gaskets, under deposits, etc. It is intense localised corrosion which occurs within a crevice or any area that is shielded from the bulk environment. Solutions within a crevice are similar to solutions within a pit in that they are highly concentrated & acidic. Also, just under the deposit or crevice, oxygen content in the solution is low, whereas, outside of the crevice in the bulk solution, oxygen content is higher. This sets up a galvanic cell, which is anodic under the deposit or in the crevice and cathodic outside, causing oxygen concentration cell corrosion. If under the crevice chloride containing environment, pH drops and chloride concentrate, this can cause accelerate corrosion and is autocatalytic.

Alloys that depend on oxide films for protection (e.g. SS & Al) are highly susceptible to crevice attack because the films are destroyed by high chloride ion concentration and low pH. This is also true of protective films induced by anodic inhibitors.

The most likely places for stress corrosion cracking to be initiated are crevices or areas where flow of water is restricted. This is due to the build up of corrodent concentration in these areas. Chloride is the main contributor to this type of attack on stainless steel (SS). If crevices cannot be avoided, sufficient circulation of environmental water and uniform composition need to be provided.

Oxygen Attack / Pitting:

Oxygen attack is an electro-chemical process as below:

Anode	Fe	$\rightarrow Fe^{2+}$	$+ \quad 2e^-$
Cathode	$\frac{1}{2} O_2 + H_2O + 2e^-$	$\rightarrow OH^-$	
Overall	$Fe + 1/2O_2 + H_2O$	$\rightarrow Fe(OH)_2$	

The reaction is aggravated at high temperature even with small content of oxygen. It is identified by well defined pits or a very pock marked surface as shown in figure 12 (d). The pits vary in shape, but are characterised by sharp edges at the surface active oxygen pits are distinguished by a reddish brown oxide cap (tubercle). Removal of this cap exposes black iron oxide within the pit. The pits can penetrate deep into the metal causing rapid failure and iron oxides thus generated can also produce iron deposits in the boiler.

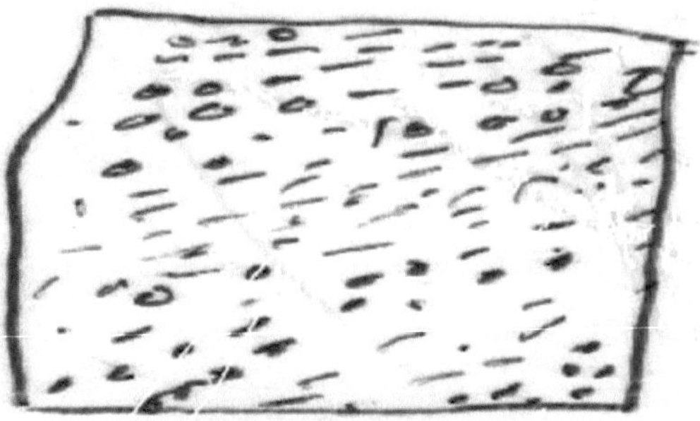

Oxygen Pitting
FIGURE 12 (d)

Oxygen attack / pitting may be highly localised or may cover an extensive area. This is generally visible in feed water pipe, feed water distribution holes, at the steam drum water line, downcomer tubes, economiser, condensate lines, etc. Unlike crevice corrosion, pitting corrosion can occur without crevices being present on themetal surfaces, but like crevice corrosion, pitting is caused by specific corrosion ions eg. chloride, etc. The propagation of pitting and crevice corrosion is the same, once the two forms of corrosion initiate, further with galvanic relationship with bottom of the pit becoming anodic and the area surrounding cathodic.

Patterns of 'pitting' by chemicals and microorganisms are shown in figures 12 (e) & 12 (f) respectively. Pitting by chemicals generally occur at low flow condition/stagnation. Pitting by microorganisms is mostly of round types & scattered. Sulphur reducing bacteria gives also black FeS precipitate. Chloride pitting of stainless steel and pitting of aluminium brass condenser tubes are also generally found in the form of holes and pits.

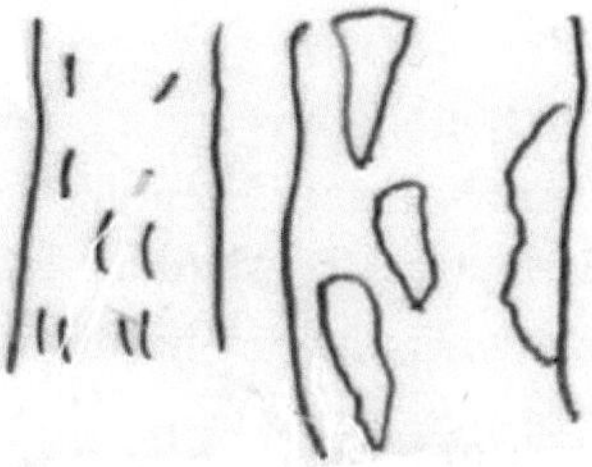

Pitting by Chemicals

FIGURE 12 (e)

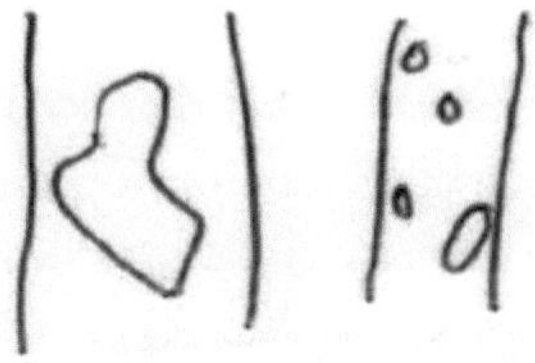

Pitting by Micro organisms

FIGURE 12 (f)

Fatigue Cracking:

This has been found due to: (a) repeated cyclic stress (b) cyclic stress in corrosive environment (c) frequent breakdown of protective magnetite film due to cyclic stress (d) repeated starts ups and shutdowns.

The metal failure occurs at the point of the highest concentration of cyclic stress, e.g

1. **In boiler**
   a. Support brackets
   b. When start up & shutdown are frequent - the rolled in tubes
   c. Horizontal tube runs due to steam blanketing
   d. Waterwall tubes-due to frequent prolonged lower header blowdown
2. **In Deaerator: near welds and heat affected zones**

   Fatigue failure on fire side of the boiler tubes have also been noted, in which, mode of failure is a crack that occurs circumferentially at the pivot point of the vibrating tube (due to air/gas flow) and may result in a complete separation of tube, almost as if, it had been cut with a 'saw'.

   Fatigue corrosion occurs in rotating parts, such as pump shafts. Pitting corrosion usually occurs on the surface, which in turn produces stress raised areas. The cyclic stress further causes accelerated fatigue cracking. Fatigue striations (beach marks) are typically observed on the fracture surface and are telltale signs of corrosion fatigue. Fatigue corrosion can be prevented by using higher strength alloys or reducing stress.

Inter Granular Corrosion:

It is a localised attack that occurs at metal grain boundaries - most prevalent in SS, which have been improperly heat treated. In these metals, the grain boundary area is depleted in chromium, and is therefore less resistant to corrosion. Intergranular corrosion occurs in specific alloys when they are heated into their sensitisation range, usually during welding or heat treatment. When alloys like some of the stainless steels are heated between 425 to 870 ^{0}C, chromium carbide precipitates out at the grain boundaries, resulting to chromium depleted regions next to the carbides and can affect the passivity of the grain boundary area. In specific environments, such as nitric acid or high temperature water, dissolution of the low chromium area can occur. The grains appear as a sugary surface, which is easily rubbed away when brushed with a probe. Intergranular corrosion of SS and nickel alloys can be avoided by using low carbon alloys by adding carbide formers such as titanium or columbium (niobium) or by stabilising anneals.

Corrosion fatigue/fatigue by inter crystalline cracking is also a predominant type of cracking across the grain structure of the metal. The cracks are often accompanied by a number of pits adjacent to or in line with the cracking - another specific indicator of the mechanical stress imposed.

Hydrogen Embrittlement:

It usually occurs in a boiler working at or above 1500 psi boiler, and is mainly due to atomic hydrogen forming at boiler tube surface as a result of corrosion process. Hydrogen permeates the tube metal where it can react with iron carbides to form methane gas, or with other hydrogen atoms to form hydrogen gas. These gases evolve predominantly along grain boundaries of the metal. The resulting increase in pressure leads to metal failure. Slow & steady diffusion of ferrous iron through the magnetite layer is also believed to give formation of hydrogen, as a result of corrosion reaction in the absence of dissolved oxygen.

The initial surface corrosion that produces hydrogen usually occurs beneath a hard dense scale. Acidic contamination or localised low pH excursions are normally required to generate atomic hydrogen. In high purity system – raw water in-leakage (condenser tube leakage) lowers boiler water pH, when $Mg(OH)_2$ precipitates, resulting in corrosion and formation of atomic hydrogen leading to initiation of hydrogen attack.

Hydrogen damage is mainly due to Cl / SO_4 (includes passage from DM plant). With this damage, full thickness chunk of metal is blown out of tube wall; ruptures are violent, sudden and disastrous. Failed section of tube is cracked in the intergranular mode, but usually retains the original thickness of tube.

Though there is no correlation between hydrogen evolution & hydrogen damage and the first indication of corrosion due to hydrogen is generally known after a tube failure analysis, measurement of hydrogen in steam is useful atleast qualitatively as a measure of initial boiler corrosion. However, this has also to be viewed in the context that the hydrogen can enter the boiler system from following sources to be found in dissolved state:

a. Oxidation of boiler steel.
b. Decomposition of hydrazine.
c. Decomposition of corrosion inhibitors like morpholine.
d. Decomposition of organic contaminants of feed water at high pressure & temperature.
e. If the feed water is heated from a gas externally which contains hydrogen. Hydrogen will diffuse through the boiler tube wall resulting in dissolved hydrogen levels high.

Acid Corrosion:

In boiler and feed water system, acid attack can take the form of general thinning or it can be localised at areas of high stress such as drum baffles, 'U' bolts, across nuts & tube ends. Causes of this type of corrosion are:

a. Leak from DM plant
b. Leak from condenser
c. Acid ions from conditioning chemicals
d. Acid cleaning

CO_2 Corrosion:

It is caused due to low pH conditions created by carbonic acids in the system and cause a generalised loss of metal rather than localised pitting as observed by oxygen corrosion. Metal destruction spread over relatively wide area, resulting in thinning. Pipe walls are thinned particularly in the bottom of pipe, as shown in Figure. 12 (g). Thinning often leads to failures especially at threaded sections.

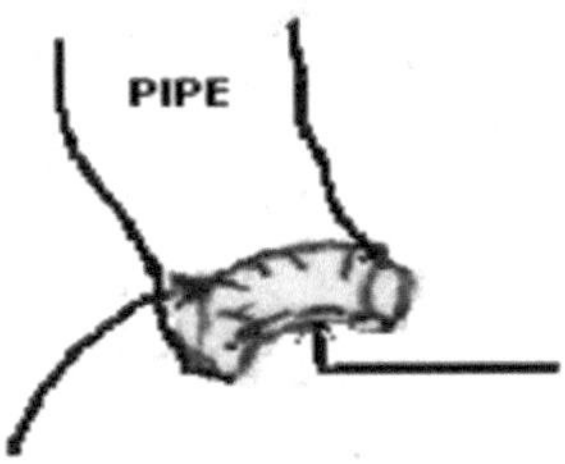

GLOBULE TYPE

FIGURE 12 (g)

Fretting Corrosion:

This is a type of corrosion, which occurs where parts are tightly fitted but slight movement can take place. Where two dissimilar metals are tightly fitted, it is usually the harder of the two (not always the case), which suffers fretting.

Erosion Corrosion:

It is an attack due to relative movement between fluid and metal like condenser tubes, pump impellor, turbine blades, etc. It is the increase in the rate of metal deterioration from abrasive effects. It can be identified by grooves & rounded holes, which usually are smooth & have a directional pattern. Erosion-corrosion is increased by high water velocities & suspended solids. It is often localised at areas where water changes direction.

Cavitation (damage due to formation and collapse of bubbles in high velocity turbines, propellers, etc.) is a form of erosion-corrosion. Its appearance is similar to closely spaced pits, although the surface is usually rough.

Selective Leaching/Dealloying:

In this case, one element usually the least noble is selectively removed by the corrosive environment leaving a mechanically weakened zone. A typical example is dealloying of brass in steam and water environments particularly in the presence ammonia. It is termed 'dezincification' where zinc is selectively removed and copper is plated back on the metal surface. The conditions that promote the pitting of steel also promote the pitting of brass, which in cooling systems usually occurs by 'dezincification'. Low pH condition (< 6.0) and high free chlorine residuals (>1.0 ppm) are particularly aggressive in producing dezincification. The dezincification resistance varies with the alloy e.g. 70/30 brass is less resistant than admiralty brass (70/30 brass +1% tin), which is less resistant than inhibited (viz., admiralty plus small amount As, Sb or P).

Steam Side Burning:

This is a chemical reaction between steam & metal caused by excessive heat input or poor circulation resulting in insufficient flow to cool the tubes, under which conditions an insulating SHS film develops. Once tube metal temperature reaches 400 ^{0}C in boiler tubes & 510 ^{0}C in superheater tubes (assuming low alloy steel metal), the rate of oxidation increases dramatically. This oxidation occurs repeatedly and consumes base metal.

This problem is most frequently encountered in superheaters and in horizontal steam generating tubes heated from the top.

Other Corrosions:

a. Acid phosphate corrosion: caused by high phosphate with pH decrease, as generally noted in boiler. Reverse pH excursions (i.e. phosphate low & high pH) is indication of 'on set' of acid phosphate corrosion, which can increase the corrosion rate rapidly up to 60 gm/ft^2, whereas usual phosphate hideout can give less than 5 gm/ft^2. However, there is no possibility of acid phosphate corrosion if:

   1. Departure from nucleate boiling is only for 1 - 2 hours
   2. PO_4 content is kept below the prescribed level as per graph
   3. Na / PO_4 is kept to prescribed ratio
   4. Magnetite layer formation is intact with 4 gm/m2

5. Low level of Cl & SiO_2 are kept
6. Saturated pressure of 2800 psi & heat flux up to 200,000 Btu/ft^2 are kept

b) Scale deposit induced overheating and corrosion. Constant bubble formation at the boiler tube wall with nucleate boiling prevents water from wetting the tube wall. Thus, corrosion product oxides & insoluble salts present in water are largely prevented to reach the tube wall to form deposits. As heat fluxes & operating pressure increase, so does the concentration factor between the bulk water & the surfaces of boiler tubes. Start ups introduce additional impurities in circuit. Attention should be given to horizontal & slanted boiler tubes including roof and nose tubes where steam-water stratification can lead to steam blanketing & impurity concentration & corrosion. Such a situation has occurred in fluidised bed boilers in the horizontal bed tubes. Boiler tube heat flux and metal temperatures on the tube internal surface can be measured.

c) The gradual stretching of metal under temperature & stress. For a given stress, the rate of creep increases with temperature. Long term degradation of metal takes place due to metal creep, change in microstructure, corrosion, erosion & metal fatigue. The extent of degradation is however based upon original properties of material, the service life of components & parametres during operation.

Limited creep rupture strength on 1% Cr, ½ Mo (AISI 304 HSS) is noted. However 304 H has better creep rupture strength.

d) Operation: On maximum design values i.e., maximum heat flux, circulation, mass flow through water wall tubes, quality, steam blanketing, etc. may change boiling pattern from nucleate type-causing corrosion. Also operation with high concentration of Fe & Cu in feed water lead to initiate boiler corrosion & other problems.

e) Flow accelerated corrosion: Low pH (at operating temperature) & excessive concentration of oxygen scavengers in feed water can lead to flow accelerated corrosion of carbon steel & high concentration of iron. In economiser, apart from pitting & corrosion fatigue, flow accelerated corrosion is found common.

f) Spalling: Rust deposits on bends of superheated tube cannot be blown off or carried away, as pressure drop across ends are very low. This causes blockage to steam flow, overheating & bursting. This is superheater steam side oxide cracking (called 'spalling')

g) Excess of acetic acid, m-phthallic acid found in condensate coming through fibre reinforced plastic (FRP) lining of hot feed water tanks, cause corrosion to LP rotors of turbine.

h) Exfoliation corrosion: This is a localised type of corrosion, in which loose flaky corrosion layers are formed on metal surface, which can be easily removed even by low velocity flow of water/steam, causing the bare metal again forming another series of flaky layers.

Mechanical / Operational conditions affecting corrosion:

How to minimise

a. Selection of corrosion resistant metals

b. Reduction of mechanical stress wherever possible (viz., use of proper welding procedures & stress relieving welds)

c. Minimisation of thermal & mechanical stresses during the operation

d. Operation within design load specification, without over firing along with proper start up & shutdown procedure

e. Maintenance of clean system, including use of high purity feed water, effective & closely controlled chemical treatment and acid cleaning whenever required.

f. Design disciplines specific to turbine: The design disciplines, which affect turbine corrosion, can be separated into 4 parts:

   1. Mechanical design (stresses, stress concentration)
   2. Heat transfer (surface temperature, heated crevices)
   3. Flow (moisture velocity, location of salt zone, stagnation temperature, interaction of shock wave with Wilson line)
   4. Physical shape (crevices, obstacles to flow, surface finish)

These four points (which interact) can produce undesirable stresses & impurity concentrations. In addition, some combination of material in contact, which are also chosen by designs, can produce galvanic corrosion.

Higher yield strength materials are more susceptible to stress corrosion & have lower fracture toughness. Hence, adding more materials does not reduce stress because the centrifugal force also increases.

Light materials have more corrosion design allowance & other corrosion related requirements, which is chosen to mitigate the problem of more impure steam problems.

Pronounced thermodynamic effect: It is frequently found that metal surface temperature is slightly above the saturation temperature of steam. It happens particularly for low pressure fossil turbine usually on L-1R blades. It is one of the reasons why fossil L-1R blade has higher failure rate than any other blade row.

Another reason for the high L-1R blade failure rate could be effects of transonic flow, vibratory stresses & changing temperature, resulting from interaction of the shock wave with the Wilson line (periodic destruction of Wilson line). It is therefore advisable to consider local surface temperature also, both in design & failure analysis. The surface may be hot because of heat transfer through the metal or because of the stagnation temperature effect (zero flow velocity at the surface and change of kinetic energy of steam into heat).

Individual Functions of Equipments in Feed Water-Steam Cycle

Condenser:

It is equipment which condenses steam after passing through turbine and forms the greater part of the boiler feed water.

The turbine exhaust steam is expanded to condensate pressure usually below the atmospheric pressure (vacuum). A typical 500 MW unit generally needs more than 20,000 m^2 of heat exchanger surface to affect the condensation of the turbine exhaust steam. The surface is generated in the form of bunch of tubes, inside which cooling water in the form of raw water at ambient temperature is continuously flowing for extraction of heat from the steam.

Hence, due to low temperature gradients involved, the heat exchanger tubes must be as thin as possible and without any damage or clogging with deposits to obstruct the flow of cooling water, so that the tubes can remain an effective barrier between the impure cooling water. The absolute pressure in the steam space at the top of a condenser is called 'condensate back pressure', which play an effective role in MW generation. Hence, there is arrangement of vacuum deloading, which protects the turbine against falling condenser vacuum, by an oil operated automatic devices which reduces the load on turbine. The vacuum loss could be due to several reasons, chief of which is air in-leakage to condenser, accumulation of deposits in condenser cooling water tubes, condenser leak, etc.

In thermal power stations, generally surface condensers as above are used, which serve dual purpose, viz., they permit low back pressure maximising plant efficiency and they recover quality feed water.

The tube materials could be of admiralty, aluminium brass, 4548 aluminium bronze, 90/10 cupro nickel, 70/30 cupro nickel, stainless steel and titanium as per demand by circulating cooling water quality around the tubes.

Feed Water Heaters:

These are heat exchangers for pre heating feed water before it passes to the boiler, usually by means of steam bleed from turbine.

In large units, these are in the form of LP heaters & HP heaters in respect of turbine bleed steam (steam from respective pressure zones of turbine).

Deaerator:

It is an important equipment for removing the dissolved gases (especially oxygen) from feed water normally involving injection of steam. A deaerator removing significant amounts of oxygen out of water to give very low oxygen residuals consists essentially of two zones - one heating the water and the other scrubbing out the last traces of oxygen as shown in the flow scheme as below.

Usually the deaerator works below 10 kg/cm^2 pressure, but for control purposes, it is best to operate at pressure above 1.15 kg/cm^2 (104 ^0C). The unit is complete with a system which ensures continuous withdrawal of incondensible gases mainly O_2 & CO_2. The deaerator also provides feed water storage and proper suction conditions for boiler feed pumps. For perfect deaerator, the amount of steam entering the vessel is closely controlled, to maintain a constant absolute pressure in the vessel of predetermined figure to give a certain outlet temperature. Availability of a small amount of carrier steam highly enriched in O_2, which it vents to atmosphere, demands for more steam in scrubbing zone, and the design must take special care to prevent this enriched vapour from contaminating the scrubbing zone. A deaerator fed with condensate which contains little O_2, does not handle enough mass of O_2 to give real problem. But where the feed is make-up water, which can contain up to 14 ppm of dissolved O_2, then the vent steam O_2 can approach atmospheric concentration to cause trouble.

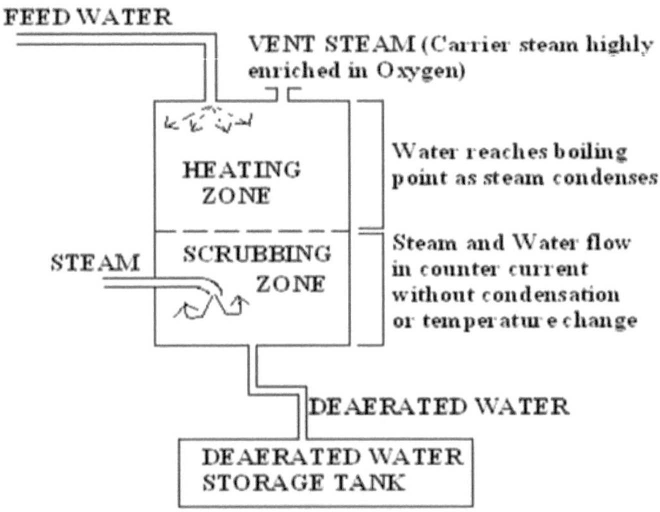

Hence, it is very much necessary to achieve proper mass transfer with controlled steam & water flow and avoid recontamination of vent steam to scrubbing zone or to deaerated water. The plate & trays type of deaerator are generally the most acceptable choice at thermal power station. Water is sprayed at the top of the dome & sprays forming heating zone. By the time water reaches the top tray, it is just about boiling. It then cascades over perforated doughnut and plain trays which serve to hold up the water flow and redivided it, in order to provide the time & interface area needed to remove traces of oxygen.

However, even with good design the plates & trays are points of potential weakness. Steam in good amount is made available either from superheated steam or turbine exhaust steam to achieve very low partial pressure of oxygen.

Economisers:

Economisers are a part of boiler used to improve boiler efficiency by extracting heat from flue gases discharged from the fire side of a boiler, and is the first area after the deaerator to be exposed to increased heat. Generally economisers are arranged for downward flow of gases and upward flow of water in the tubes, which form the heating surface. In TPS, generally non-steaming economisers are used in view of steaming economisers being more sensitive to deposition & erosion at tube bends. Also steaming in the economiser is characterised by a rapid rise in the steam drum water level or water hammer when feeding water. Hence, sufficient flow through economiser needs to be maintained to prevent the generation of steam in economiser.

Boiler Drum:

A vessel usually fabricated by fusion welding or solid forged. It provides a reserve of water for circulation and a space in which steam and water are separated. The basic function of this drum is to provide a large volume to reduce the upward velocity of steam/water mixture, in view of the moisture droplets, being denser than steam to fall back to the water surface. Drum internals e.g. baffles, plates, screen & centrifugal separators help reduction in drum size requirements & provides a means for elimination of the finally divided moistures spray, which otherwise could not be removed. For elimination of moisture in steam, the screen drier drain pipes usually terminate below the water level of the drum & feed water inlet nozzles are aimed below horizontal.

A boiler in which no water recirculation takes place is called 'once through boiler' and does not contain water/steam drum. The other function of the drum is to arrest unwanted impurities of water in the form of sludge and remove by blowdown to drain. However, blowdown operation is not done to lose sight of water in the gauge glass. Also, use of bottom blowdown valves (applicable to all type boilers) should be restricted to periods of moderate steaming rates & never at high loads.

Superheater:

This is a part of boiler used for adding superheat to saturated steam, which is generated in boiler having no liquid phase and the temperature of which is equal to the saturation temperature corresponding to its pressure. The complete superheated section of a boiler may contain several separate superheaters. A superheater in which tubes are arranged vertically is called 'pendant superheater'.

Reheater:

It is that section of boiler in which exhaust steam from the high pressure cylinder of turbine is reheated usually to its original temperature before entering the intermediate pressure cylinder of the turbine for further work.

Turbine:

It is a heat engine used to convert the heat energy in steam to mechanical energy. The superheated steam is passed through the turbine cylinder, which is turbine casing assembly housing the fixed/moving blades & the rotor, in such a way that cause a very high speed of the cylinder. This when coupled with the moving cylinder of the generator, generates electricity. From site to site different types of turbines are found, but have generally the following components/facilities in common:

a. Barring gear: An equipment for rotating a turbogenerator shaft on speed, usually motor driven.
b. Governor: A device which controls the speed or load of a machine.
c. Excitor: A small generator for supplying the direct current required for the rotor winding of a large generator.
d. Diaphragm blade: It is one of the stationary blades which is fitted to the casing of a turbine.
e. Gland steam: Arrangement to provide steam used to prevent air from entering a turbine cylinder between the turbine shaft and casing.
f. Differential expansion measurement: A device to know the quantum of expansion between rotor shaft and casing.

g. Vibration metre: There should be limit of vibration on turbine at its different speed. The more will either indicate deposits on turbine blades or some problems including of its bearing, etc.

h. Resonance indicator: This is the condition to know in which a natural frequency of vibration (or oscillation in electrical circuits) reinforces an applied vibration. Thus, in turbine maximum vibration occurs at certain speeds due to resonance at these speeds.

i. Synchronism facility: A procedure followed when connecting an incoming alternator to the bus bars, so that the frequency & phase relationship of the two sources of power are identical.

j. Bus bars: A continuous set of separate copper conductors extending the full length of a switch board, to which all the circuits on the switch board are connected individually, by some form of circuit breaker or switch.

Air Heater:

This is heat exchanger in the path of boiler gases after economiser, so that air from combustion absorbs heat from gases before they enter the chimney stack.

Metallurgical Perspective:

Selection of materials for thermal power plants are generally undertaken to take care of long durable operation, against the following:

a. Erosion

b. Corrosion

    i. Oxidation

    ii. Stress corrosion or intergranular

    iii. Pitting corrosion

    c. Weldability

In view of these, the following metals when combined with metallic iron to form alloy steel are found helpful as noted below:

Cr, Al, Si & Cu: These tend to produce adherent oxide films on the surface of steel, which increases its resistance to corrosion at elevated temperature. Also, pitting or cavitation can be avoided by annealing the material at 650 ^{0}C.

Silicon and aluminium improve oxidation resistance. Copper is added to increase resistance to reducing acids, particularly sulphuric acid. Copper is also added in large amounts (3 - 4%) to produce an alloy that has a lower work – hardening rate and can then be readily formed such as in cold heading operations for fasteners.

Silicon is a ferrite former, while carbon, nitrogen, manganese, copper and cobalt act like nickel to promote austenite.

Cr: Stainless steels have one fundamental aspect that they all contain the element chromium. At about 12%, this element retards corrosion by spontaneously forming a stable, transparent passive film. Even at 10%, Cr gives definite benefit to corrosion resistance to superheated steam and wear resistance also. Low Cr alloys are resistant to ferrous chloride attack.

Depending on the levels of alloying elements, viz., Cr, Ni, Mo and others, stainless steel can be ferritic, martensitic, austenitic or duplex in structure, affecting the metallurgical structure, physical properties and corrosion resistance of thestainless steels and nickel-based alloys. As per phase diagram for the binary chromium - iron alloy, at a given carbon content, the so called austenite loop (gamma loop) occurs at about 11 - 13% Cr. This can be extended to about 16 - 18% Cr, if other austenite-forming elements are added. Of particular note are the effects of carbon, nitrogen and nickel, which expand the range over which austenite is stable. If a stainless steel can pass through the gamma loop during the process of heatingand cooling, it undergoes the ferrite-austenite-martensite transformation and is called martensitic. Typically, such a stainless steel is magnetic, like iron, and can be hardened by heat treatment, and on the other hand, an alloy of about 17% Cr with little or no austenite forming elements being outside the gamma loop, retains its ferritic structure and cannot be hardened by heat treatment, and is also magnetic, because of the ferrite structure, thus called ferritic stainless steel. Chromium is thus called a ferrite former, because it suppresses the transformation of ferrite to austenite. For very low carbon alloys, more than 13% Cr is needed to remain ferritic.

Ni: This is for high strength, toughness and ductility with corresponding ease in fabrication, manufacture and welding. Corrosion resistance to acids, particularly reducing acids is also enhanced. Nickel also increases the alloy's ability to maintain passive films and helps in resisting corrosion in caustic environments. Since nickel promotes the stability of the austenite phase, it is called an austenitiser.

Ni-Cr: The two are good combination for case hardening elements & also for steel resisting corrosion and oxidation at elevated temperature. If iron is allowed with about 18% chromium and 8% nickel, the nickel opens the range of austenite to low temperatures even at sub zero temperatures, with ferrite still forming but very sluggishly.

The austenitic alloy is characteristically non-magnetic as well as non-hardenable by heat treatment.

Mn: When 1.65 to 2.1%, termed as alloy steel, contributes strength, increases hardenability.

Mo: Increases the passive film strength and improves pitting and crevice corrosion resistance, especially in the presence of chloride ions as in halogens salts or sea water. Molybdenum also provides increased resistance to chloride stress corrosion cracking. It increases the creep strength (due to dispersion of fine carbides-cf., Mo_2C, $(FeMo)_6$ C at high temperature). 0.5% Mo gives supplement strength at boiler temperatures above 400 ^{0}C. It is a powerful resistant to some of the microstructural changes, viz.,

a. Brittleness: Reduction in impact strength of Ni-Cr steel at 250 - 450 ^{0}C. Ni-Cr-Mo steels are used for turbine rotors

b. Graphatisation: Breakdown of carbide in a steel into graphite at 425 ^{0}C in carbon steel. Steels containing 0.4% to 0.6% Mo and 0.4 to 1.5% Cr are suggested for 425 - 540 ^{0}C.

V: It is a powerful deoxidiser. It slows down tempering at 500 - 600 ^{0}C - has superior creep resistance than Mo.

Stellite material (Co-Cr-W): From this lining is done on material where erosion problems are more, viz., from high velocity steam, fire gas and also against pulverising coal.

Titanium & columbium (niobium): These are included to preferentially combine with carbon and nitrogen to form carbides and nitrides. This improves high temperature strength properties and precludes the formation of chromium carbides which would leave areas adjacent to the grain boundaries depleted in chromium and susceptible to intergranular attack. Specific heat treatments are required to maximise the effect of the added element for stabilised stainless steels. Columbium also imparts some high temperature creep rupture strength.

Cerium, Yttrium and Lanthanum: These rare earth elements are added to increase oxidation resistance and allow lower levels of chromium and nickel to oxidation resistance to alloys with higher levels of alloying elements.

S, Se and Pb: These improve machinability, but tend to reduce corrosion resistance under severe conditions.

C and N: Carbon and nitrogen are interstitial alloying elements and improve both room temperature and elevated temperature mechanical properties. The improvements result in increased yield strength tensile strength and creep rupture values. In austenitic and duplex stainless steels, nitrogen increases pitting resistance and lessens the chance that the intermetallic phase, sigma, will precipitate at elevated temperature or during welding. In view of the above, therefore, the following steels are chosen against the undernoted equipments:

Boiler tubes & drums: These are generally made from low carbon or carbon molybdenum steel, and fabricated from bent plate or forgings, fusion welded through & suitably stress relieved.

Superheater tubes & high pressure piping: For these Cr-Mo alloys, in which carbon not greater than 0.15% are there, are used as below:

	C	Cr	Mo	Ni	Columbium
Up to 595 ^{0}C	0.15	0.2 - 6	0.5 - 1	x	X
Above 595 ^{0}C	x	18	x	8	trace

These have suitable creep stress, rupture strength and resistance to corrosion and oxidation at elevated temperature. Creep is generally 1% in one lac hours.

Turbine HP cylinder:

Alloy steel with following composition is generally in use:

400 - 440 ^{0}C	C + 1/2 Mo steel
440 - 565 ^{0}C	C + 1/2 Mo + 1% Cr + V steel
565 ^{0}C & above	C + 2% Cr + 1% Mo steel

Turbine rotors:

Cr-Ni-Mo steel up to 565 ^{0}C are generally in use, which takes care against resistance to corrosion wear, temperature brittleness & mechanical properties.

Turbine blading:

Up to 485 ^{0}C	1% C, 12% Cr
Above 485 ^{0}C	18/8 Austinitic steel or 12% Cr- Mo - W

These have high temperature strength and stabilised microstructure main criteria.

Control & throttle valves:

For this same material as high pressure pipings for steam is used.

Valve seatings:

For these satellites material are generally in use, which are hard enough (cf: tungsten carbide) to meet hazard of erosion.

Chapter C

COOLING WATER CHEMISTRY

(Describes impact of cooling water in respect of fouling, scaling and corrosion and various means to contain them, highlighting along with monitoring & evaluation of treatment programme of the cooling water including sea water dealing in detail with macrofouling problems particularly from 'Barnacles')

Cooling Water Chemistry

There is huge water reqirement for cooling systems for proper functions of a TPS with specific qualities to avoid problems of deposits, scaling, corrosion, etc. in the cooling systems. Areas of cooling in a TPS are generally condenser tubes, various auxiliaries including various high pressure pumps, lubricating oil cooler, air cooler and stator water cooling, etc. In view of condenser cooling requiring a vast amount of water for cooling the whole steam as used for rotating the turbine, full fledged processing of cooling water is found often uneconomical and a TPS has to depend on the usual source water (raw water) with little or no processing, but with suitable chemical dosings to minimise its adverse effects. Since in other areas, the cooling water requirements are comparatively very small, the source water is often needed for partial to complete processing as per the site specific cooling systems and the availability of source raw water.

Types of cooling:

A cooling system for any heat exchanger can be found in mostly three ways:

1. Once through type.
2. Open recirculating type.
3. Closed recirculating type. However, in view of general scarcity of source water at site, in an attempt to conserve water, recirculating type of cooling are generally adopted nowadays.

The three types of cooling are defined as below:

Once Through:

Here is single pass flow through the heat exchanger to return the whole flow to the receiving body (source water). In view of the large quantity involved as in case of condenser tube cooling, there is fear to contaminate the receiving body including with heat, which may not be permitted under pollution control norms. Hence, search has to be made for effective and economic ways to treat water before putting it down again to the receiving body. Choice of water in this case is therefore generally raw water from the source, but after treatment against prevention of flowing coarse debris and accumulation of micro/macroorganisms into the cooling system. Under this category of cooling, usually condenser tube cooling is in practice at TPS, where raw water is found in abundance.

Open Recirculating:

This is a system in which heat absorbed from the process is dissipitated in cooling tower, spray ponds or evaporative condensers. Here, water is in reuse with system open to air in cooling tower & then through duct or channel back to cooling circuit. There is some loss of water here due to evaporation & drift loss for which minerals as present in the source water are found concentrated and the cooling system water needed to be given blowdown with suitable make-up with source water (usually in the range of 2 to 4% of recirculation rate which includes loss of water due to evaporation and drift) to limit the concentration of dissolved salts to a fixed cycle of concentration (ratio of hardness or TDS or silica in circulating water to the same in make-up water). Choice of water for this system is either raw or clarified water, which is needed only as make-up quantity. Still the system has greatest potential for all types of problems of deposits, corrosion & micro/macro biological problems contributed from make-up water, air & from the system itself. Hence, there is greatest need for solution/treatment. The situation becomes aggravating, if same water has to be used either for drinking or even discharging into receiving body, etc. In most of the TPS in the country or abroad this type of cooling are in practice for their condenser cooling system.

Closed Recirculating:

This is a cooling system within a cooling system where the heat is exchanged with another fluid. The cooling water is completely confined within system with absorbed heat dissipitated by air cooling. Loss of water is only through leakage and hence very little water is required for make-up (usually 0.1 to 0.2%) keeping the mineral content of cooling water virtually constant. A very high quality of water including DM water could therefore be used here, which minimises potential water problems. Yet, corrosion remains here as most serious problem, in view of corrosion products not removed and foul the system. Hence, suitable corrosion inhibitor is required to deal with this situation.

At few sites, softened water is also used in place of DM water. In TPS, this type of cooling is generally adopted for auxiliaries cooling circuit. Available sources of water with its properties causing concerns for smooth function of a TPS can be viewed as below:

Sources of Water

GROUND	SURFACE	SEA
Wells, borewells, springs, etc.	Lakes, ponds, rivers, reservoirs, canals, etc.	Sea, back sea, estuarine brackish water, etc.

1. **COARSE DEBRIS MATTERS** - Vegetation, fish, rubbish cause choking
2. **FINE SUSPENDED** - Sand, clay, dust, corrosion products cause choking erosion, deposits cause induced corrosion
3. **MICROORGANISM** - Bacteria including S & Fe cause choking and corrosion
4. **MACROORGANISM** - Shellfish, snails, clams, mussels, barnacles, algae, fungi, etc. cause choking
5. **Free CO_2** - CO_2 in excess of equilibrium gives corrosion & less $CaCO_3$
6. **Fe & Mn** - Cause fouling & are nutrient for bacteria

7. CALCIUM SULPHATE - High amount gives scaling & fouling
8. CHLORIDE, SULPHATE, OTHER DISSOLVED SALTS GASES, pH, etc.

Potentially corrosive ingreditents

Coagulants, detergents, sewage, corrosion products, inhibitor reactants, etc. can also cause severe fouling, if there is a big gap between two successive cleaning or operation is being done at higher temperature & high heat transfer rate.

Though the problems of cooling water in the cooling cycle vary from water to water, the following main problems can be addressed with under noted control measures:

Fouling:

The foulants are generally mud & silt, natural organics, dust & dirt, vegetation, microorganisms, macroorganisms, oil, process contaminants including corrosion products & chemicals used, iron & manganese oxides, etc.

These can be controlled by suitable processes like screening, chlorination, surfactants, dispersants, side stream filtration, sponge ball cleaning, etc.

Scaling:

These are generally water soluble deposits caused due to inherent quality of water in view of presence of calcium bicarbonate/carbonate, calcium sulphate, calcium phosphate, magnesium silicate, carbon dioxide, etc.

Control measures for these are by adopting processes like using softened make-up water, controlling deposit tendencies by adding chemicals like antiscalant, dispersants, etc. by keeping attention on Langlier, Ryznar, Puckorious index, etc.

Corrosion:

Ingredients responsible for corrosion in the cooling water circuit having mixed metallurgy but dominated by mild steel are dissolved & suspended solids, dissolved gases like CO_2, H_2S, NH_3, O_2, Cl_2, etc, in water along with degree of alkalinity or acidity of water & its velocity and temperature. Presence of bimetals and microbial growth in the cooling circuit and imbalance of chloride & sulphate of cooling water may further lead to severe corrosion.

Control measures towards these are adopting water processings like reverse osmosis, filtration, softening to the extent balancing chloride with hardness and using anti-corrosive chemicals, cathodic protection device along with applying protective coatings & protective film formation techniques by suitable chemicals.

Biological Deposits & Corrosion by Slimes:

These are caused by plant & microorganisms like algae, bacteria, etc. and mussels, clams, barnacles, etc. and the adhesive slimes generated by them.

Control processes for this is suitable pretreatment of water, chlorination, dosing of various biocides, etc. The treatment practices for the cooling water as followed in TPS, to contain the above problems can thus be elaborated below as per following heads:

Fouling:

Heat exchangers are commonly designed for fouling factor of 0.001 to 0.002. The fouling factor (resistance due to fouling) *R* is a relationship between the initial overall heat transfer coefficient (U) and the overall heat transfer coefficient during service (U). The fouling can be contributed mainly by the suspended & biological matters as noted below:

By suspended matter:

The raw water as received from the source is tried to be taken into cooling circuit via a low velocity settling tank, if possible, otherwise through bar, coarse & fine screens to arrest maximum of debris & suspended matters of the water. The bar screens have bar distances of generally 100 - 150 mm. The coarse screen is generally of 3" size and the fine one of ½" size. More fine is also adopted in the form of travelling water screen, which is able to arrest suspended particles even of size 0.4 to 0.5 mm. Further, in some power stations, there is provision for side stream filtration, figure (1 or 2), filtering a small quantity continuously to minimise the quantum of suspended matter from the cooling tower basin apart from dredging arrangement of the cooling water basin/sump to remove the settled solids in the basin/sump, which may otherwise be circulated to foul the cooling system.

Cross sections of the cooling water pipes are also such that velocities of approx. 2 m/sec occur. For operational & design reason, this value is reduced to 0.5 - 1 m/sec at some points in the prescreening system, culverts. Yet clogging of the 15 - 30 mm inner diametre condenser tubes still occurs, which could be due to defective water screen. Also, aquatic organisms pass through mesh of recirculating water screen grow further in mains and culverts.

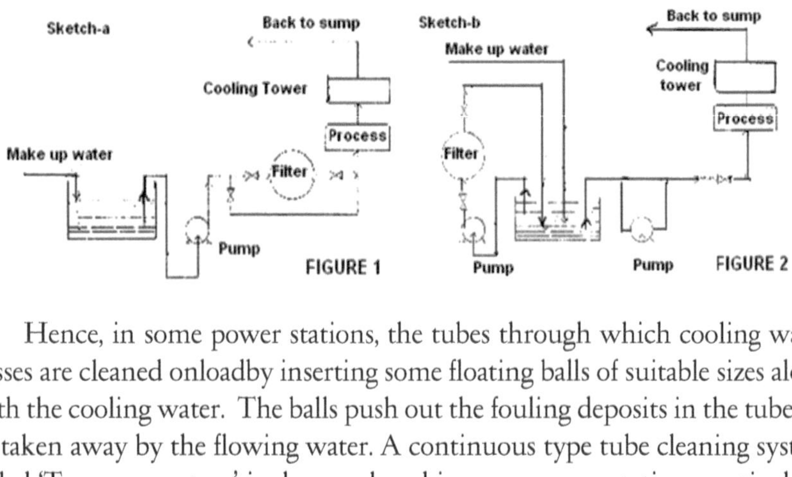

FIGURE 1 FIGURE 2

Hence, in some power stations, the tubes through which cooling water passes are cleaned onloadby inserting some floating balls of suitable sizes along with the cooling water. The balls push out the fouling deposits in the tubes to be taken away by the flowing water. A continuous type tube cleaning system called 'Taprogge system' is also employed in some power stations particularly on sea water cooling system. If the tube surfaces are still found fouled, offload cleaning either mechanically (with soft nylon brush) or chemically, is done. When the cooling water in circuit appeared to be much dirtier in comparison to the make-up water, suitable amount of blowdown from the circuit are given & replenished with the fresh make-up water.

Use of dispersant chemicals are again one of the ways to deal with the limited amount of fine suspended solids either originally present or formed as precipitation product of calcium carbonate, phosphate, etc. These chemicals keep the fine suspended matters always in dispersed phase, not allowing them to settle on to foul the tube surfaces. Addition of lignin, tannin (50 - 200 ppm) had been the old practice in this connection, which also worked as excellent food for microorganisms and reacted with chlorine & iron salts. Of late, use of low molecular weight polymers, particularly of anionic nature, in this regard is at large, including polyacrylates (4 - 5 ppm). Sludge fluidisers and some surfactants are also on use towards this. Sludge fluidiser is generally high molecular weight polymer like polyacrylamide. These flocculate/agglomerate fine suspended matter to larger high molecular weight (in the millions; and hence opposite function to polyacrylate dispersant). Its dose as 1 ppm (active) is beneficial in once through cooling system for a short time each day, but the same is not advisable to be used for cooling tower system. For industrial waste including from oil & hydrocarbons, surfactants & wetting agents are used.

By Biomass: Biofouling, in general cooling water, takes two forms:

a) Biological fouling of heat exchangers (caused by algae, bacteria and plant & animal organism), and macro biological fouling of intake and discharge channel (by inverteberate life, mussel & clams etc). Fouling of the cooling water circuit by micro / macroorganisms is also too frequent. Even the slimes generated by them are some times found so much sticky that it can entrap other suspended particles also considerably to choke the cooling water tubes. The threat is presented by uncontrolled growth of these organisms. Systems that provide the water storage & aeration are particularly notorious in this respect. A cooling tower system is an excellent example. Temperature & pH of cooling water are usually ideal for microbial growth and the presence of organic & inorganic salts ensures plenty of nutrients.

   Warm sunlight and O_2 enhance the favourable environment. Biofilm (slime) formation causes flow losses not only by physical blocking of a passage, but also through such factors as kinetic energy absorption from flowing water by its rippled surface contour (as shown in figure below), which some times are invisible to naked eye, since it is 90-95% water.

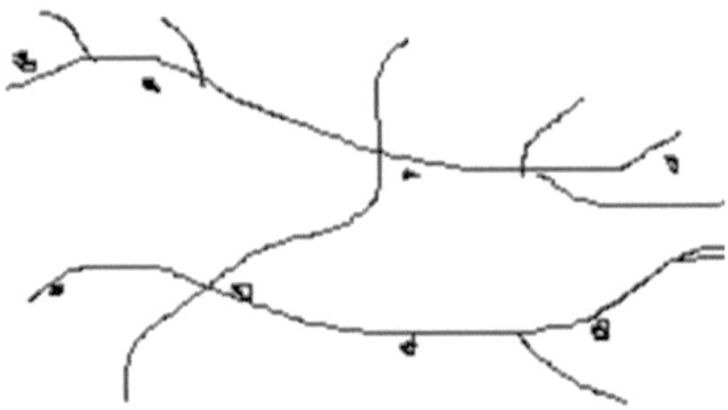

A 500 micrometre film will dry to a thickness of 10 - 25 micrometre, when water is removed. Thus, uncontrolled bacterial slime can reduce heat transfer efficiency significantly and aggravate condenser tube corrosion.

The following biocides are generally used to control the menace of these biomasses:

Oxidising biocides

1. Chlorine
2. Chlorine dioxide
3. Chlorinated bromine
4. Hydrogen peroxide
5. Sodium/calcium hypochlorite
6. Isocyanurates (active halogen compounds)
7. Ozone
8. H_2O_2
9. Bromine

Non-oxidising biocides:

1. Quaternary ammonium salts (alkaline pH range)
2. Methelene bis thiocyanate (MBT)
3. Organo tin compounds (alkaline pH range)
4. Phenate based compounds
5. Dodecyl guanidine hydrochloride
6. Triazine
7. Polymers

Amongst all, chlorine is the most familiar, cheap and effective industrial biocide. Hypochlorous acid as formed with reaction of chlorine with water easily diffuses through cell walls of micro/macroorganisms to reach cytoplasm & produce chemically stable nitrogen chlorine bond with cell proteins. Chlorine also oxidises active sites on certain coenzyme sulphydryl groups (an intermediate step in production of 'adenosine triphosphate' (ATP), which is essential aspect to microbial respiration. The optimum pH for chlorine reaction is 6.5 - 7.5. Reducing agent and nitrogeneous matter including ammonia, if present, demands for more chlorine dose, but there is danger of producing some poisonous nitro chloro compounds also.

Quaternary ammonium salts are the most common & effective among non-oxidising type biocides and can be taken along with or as replacement against chlorine. It is most effective against algae & bacteria in alkaline pH range. Surface active (+) charge on it forms electrostatic bond with (--) vely charge sites on microorganism cell walls – thus creating stresses in the wall leading to cellolysis & death by protein denaturation. But surface activity of the chemical tends to emulsify oils, weekening the cell wall bonding. However, activity is lost in presence of dirt, oil & debris, if an, y found in system and is less effective at higher temperature. Also overfeeding of this salt causes extensive foaming.

Among the organo sulphur compounds, which have been found best for fungi & bacterial slimes are methylene bis thiocyanate (MBT). It is very effective to kill desulphovibrionales species (sulphate reducing bacteria), but this does not work in presence of dissolved ferric salts and hydorlyses above pH 7.5. Bis trichloromethyl sulphone (working pH 6.5 - 8.0) & tetrahydro dimethyl thiadiazine thione (working pH > 8.0) and sodium dimethyl dithio carbamate & disodium ethylene bis dithio carbamate (working pH 7.5) have also been found effective as par with MBT. Similarly iso thiozolinone (working pH wide range) have been found as best algicide in the market, which acts against bacteria & algae by cleaving their protein bonds. Organo tin compounds are toxic for algae, mould & wood rotters, but its discharge into receiving body could be objectionable. Phenate based compounds could be seen very effective in dealing with fungi only in place of chlorine. Dithiocarbamate also is sometimes used as an alternate feeding in place of quaternary ammonium compound.

Amines & diamines (surfactants) combined with chlorinated phenolics enhance microbiocidal value. Dibromo nitrilo propionamide (DBNPA) selectively brominates/ oxidises specific substrates within the microorganism. It is sometimes categorised as an oxidising type microbiocide. Its half life falls off sharply with rising pH & temperature. DBNPA hydrolyses and is destroyed.

Chemical treatments either against fouling of suspended matter or biomass are effective if they are properly used, since all have limitations and require proper matching to the particular foulant and fouling conditions. No one chemical is known to be a universal foulant control agent. Care must also be taken to ensure that the different chemicals do not negate each others' effects or produce additional problems. For example, antifoulants & biocides that are uncompatible or interact, can lead to additional fouling. Similarly, improper application of some corrosion inhibitor or a slime control programme can cause severe fouling even in presence of a good foulant control programme.

Biocides have to be chosen in such a way that these can work at neutral pH & above as generally found in natural water and also should be effective for control of algae, fungi & bacterial slimes together. Also, to avoid immunity to bacteria to particular biocides, it is better to use atleast two types of biocides alternatively at predetermined duration.

Biocides effectiveness:

Effectiveness and characteristics of some of the biocides against fouling microorganisms (no separate biocide is listed so far, for macroorganism; the same are still on trial). Chlorine is best to deal with such larvae, but a continuous & uninterrupted application is required. Quaternary ammonium compounds & certain surfactants also have latent mortality effect under exposure for 6 - 48 hours.

Hence, it is necessary that half life of biocide (which represents time required in an hour, to dilute the activity of biocide chemical to 50%) as dosed, should be maximum as far as possible. This means holding time index (HTI) of the cooling tower basin should be sufficiently large up to 60 - 100 hours, if possible. This index is calculated by dividing 0.693 of hold up volume (m^3) by the blowdown amount (m^3/h). If the index is less, chances of microfouling will be less, but macrofouling control would be difficult.

The extent of fouling due to biomass can also be interpreted by the following table for suitable remedial measures:

Over and above, it has been seen that if in a cooling water circuit, cooling water flow is maintained althrough 3 - 5 ft/sec generally and 5 - 8 ft/sec occassionally, the heat exchanger is likely to be free from fouling. At high rates of flow, the shear stress of the water exceeds the shear strength of ordinary deposits, and they are swept away. Increasing velocity of cooling water is also therefore one of the ways to control the fouling in cooling system. Similarly by raising cooling water temperature to 40 ^{0}C many times a year may be helpful in controlling biological fouling.

Scaling

Scaling, by and large, is due to precipitation of dissolved salts present in the cooling water either due to supersaturation or any chemical reaction, and depositing on the cooling system surfaces to cause moderately hard to even very hard deposit. Some foulants including corrosion products/biological slimes, etc. when allowed to dry can also give hard incrustations to be perceived as scales.

Supersaturation

In the cooling water, certain slightly soluble salts like phosphate, carbonate, sulphate of calcium, silicate of magnesium, etc. get supersaturated on attaining their certain concentrations either due to recirculation or due to attaining certain temperature, and get the portion precipitated beyond their solubility values at the prevailing conditions, to cause scaling. Total hardness is a rough measure of scale forming properties unless both carbonate & non-carbonate hardness are known separately. Non-carbonate hardness is approx. 70 times more soluble than the carbonate ones, and hence in many cases non-carbonate hardness as much as 1200 ppm is not so serious for deposit of $CaSO_4$ in cooling cycle is mostly insoluble at pH 7 to 9. Only when all CO_2 is removed, will Ca deposit as sulphate. Safe upper limit of Ca & SO_4 concentration in absence scale inhibitor $(Ca^{2}+) x (SO_4^{2-}) = 5,00,000$. Solubility of $CaSO_4$ increases in presence of high chloride content.

Hence, when Cl > 5000 ppm, limiting Ca hardness to 700 ppm is beneficial; regardless of SO_4 levels - there will be no $CaSO_4$ precipitation. A rule of thumb for magnesium silicate precipitation will be when exceeding the limit $[Mg^{+2}]$ x $[SiO_2^{\ -}]$, as $CaCO_3$ ppm, is equal to or less than 35,000. An upper limit of SiO_2 in circulating water is 150 ppm at pH 8.5 and above. Magnesium hardness may be as low as 50 ppm. Magnesium silicate potential problem of scaling is thus at pH>8.5. Calcium phosphate is less soluble at higher pH & temperature. If Ca hardness is 500 ppm & pH 7.0, even 5 ppm of PO_4 causes scaling.

Scale Occurrence in Cooling Water System

Most common	% Solubility		Less frequent
	20 ^{0}C		100 ^{0}C
Calcium carbonate	0.005	0.002	Magnesium silicate#
Calcium sulphate	0.3	0.06	Zinc phosphate *
Calcium phosphate	0.0003	0.0002	Calcium fluoride
Ferrous hydroxide	0.0007	--	Iron carbonate
Ferric hydroxide	0.0001	--	*from corrosion control
Silica	0.005	--	chemicals
			#potential problem of scaling
			above pH 8.5

Chemical Reaction:

Apart from certain chemicals as used in the cooling circuit, which may get precipitated by reacting with each other, the main causing concern towards this is the dominant presence of calcium bicarbonate in most of the cooling water. More the presence of calcium bicarbonate, viz., and M-alkalinity, there is more likely precipitation of calcium carbonate species as decomposition product of calcium bicarbonate. This however depends on equilibrium mixture of following species in water, where CO_2 plays an important role to keep $Ca(HCO_3)_2$ in equilibrium with solid $CaCO_3$.

$$CO_2 + H_2O \rightarrow H_2CO_3 \rightarrow H^+ + HCO_3^- \rightarrow ½\,CO_2 + H_2O + CO_2$$

$$H_2O \rightarrow H^+ + OH^-$$

Thus, when water is deficient in free CO_2, it will tend to deposit some of Ca $(HCO_3)_2$ which it holds in solution, in the form of solid $CaCO_3$. Conversely, if free CO_2 is more, water will dissolve even protective coating of $CaCO_3$ or any other solid $CaCO_3$ in contact including RCC structure. This is called 'aggressive CO_2' and increases corrosion rapidly. The pH however is best to know practically in this way. Water is placed in contact with powdered chalk & pH is measured at which no longer there are deposits or dissolves $CaCO_3$. This pH is designated as pH. Hence, based on the calcium & M-alkalinity content of the cooling water along with total dissolved salts & the prevailing temperature, certain indices were framed to predict, in advance, the tendencies of water of scaling or helping corrosion to the cooling system. The most common of these indices are Langlier (saturation) Index, followed by 'Ryznar (stability) Index' and later 'Puckorious (modified stability) Index'. These are defined as below:

Langlier (saturation) Index (Lsi)	=	pHobserved - pHsaturation and, pHsaturation
	=	16.58 - log (Ca) + log (M. Alk) 0.1log (TDS) - 2.6 log (0F)

An Index value of	'zero'	indicates the water to be non-scaling
	' + '	indicates the water to have scaling tendency
	' - '	indicates the water to have corrosive tendency

Ryzner (stability) Index, (R_{si})	=	$2pH_{saturation} - pH_{observed}$ > 6, indicates scaling tendency > 6, indicates non-scaling tendency
Puckorious (modified stability) Index, (p_{si})	=	$2pH_{saturation} - pH_{equilibrium}$, Where, $pH_{equilibrium}$ is based on total alkalinity, and its value is generally taken as 1.465

Predicting indications of this, index remain same as that of Ryzner (stability) Index, as above. Scaling severity keyed to index, can further be read as below:

L si	Rsi	
3.0	3.0	extremely severe scaling
2.0	4.0	very severe
1.0	5.0	severe
0, 5	5.5	moderate
0.2	5.8	Slight
0.0	0.6	stable water

Similarly, corrosion prediction can be found as below:

-0.2	6.5	very slight corrosive tendency
-0, 5	7.0	slight corrosive tendency
-1.0	8.0	moderate corrosive tendency
-2.0	9.0	strong corrosive tendency
-3.0	10.0	very strong corrosive tendency

Langlier index fails to asses $CaCO_3$ scaling tendency in case of sea water, because of high ionic strength (conductivity) and ion pairing that characterise this water.

Prevention of Scale Formation

Scaling, in general, 15 - 30 mg/cm2/yr of deposition is expected, which cannot be taken as concern, yet it is necessary to minimise, and in no case the same should exceed the limit. The following are the ways to control the scale deposition in the cooling circuit:

1. **By blowdown & limiting COC:**

   Based on the above tendencies of scaling, which is mostly due to precipitation of calcium carbonate, out of the M-alkalinity present in almost all the cooling water, the concentration of calcium bicarbonate is tried to be kept in limit, while the cooling water is on recirculation through cooling tower, when in each cycle of recirculation, the water is progressively getting concentrated with dissolved salts Hence the need of limiting the number of such cycles arises, better termed as cycle of concentration, called as 'COC'.

As cooling tower water is concentrated and pH is increased, the tendency to precipitate scale forming salts increases. Hence, 'COC' were initially limited primarily to prevent scale formation based on the calcium bicarbonate content of the cooling water as available to site as per the table sited below. It is practically controlled by way of blowing down of the water from the cooling cycle and replenishing the shortfall with the fresh water as make-up. However, as water's TDS level increases, both corrosion and deposition tendencies increase, the deposition of salts other than calcium carbonate. It becomes progressively more difficult and expensive to inhibit corrosion as the specific conductivity approaches and exceeds 10, 000 micromhos/cm.

Also, in the light of CO_2 equilibrium posing for deposits or corrosion as described above, it can be noted that in a cooling tower function, equilibrium is never achieved partly because CO_2 content of air within the tower shell is higher than that in atmosphere, due to CO_2 released from water as it cascades down the cooling tower on each circuit. Hence, amount of CO_2 lost from water, while it remaining in the system will depend on COC. Use of clarified water as make-up is therefore helpful in this situation, replenishing loss of CO_2 to some extent (meaning thereby reduction of fouling deposit of $CaCO_3$), but care has to be taken for corrosion also, when CO_2 level could also approach higher than the equilibrium level. As already seen, the excess CO_2 from equilibrium can be removed by proper aeration.

$Ca(HCO_3)_2$, mg/l	COC
100	2
150	1.7
200	1.5
250	1.3
300	1.1

Further, to keep the COC to desired value, the amount of blowdown to be given from the cooling circuit, having cooling tower is calculated as below:

Blowdown = Evaporation Loss / COC-1

Hence, make-up water quantity = evaporation loss + blowdown + drift. The drift quantity in any cooling tower (of forced draft) is generally taken as 0.05 to 0.1% of the recirculation rate of the cooling water.

The evaporation loss from such tower (forced draft) is also calculated as below:

$$\text{Evaporation loss } (m^3/h) = \frac{\text{Recirculation rate } (m^3/h) \text{ x delta t } ^0C}{556} \text{ x f}$$

where, delta t^{0}C	Temperature difference between the inlet & outlet cooling water
and f	Tower factor, generally taken as 0.75 to 1.0, and even 0.6 for very low evaporation in cold countries. For tropical countries & conventional tower, it can be easily taken as 1.0

2. **By reducing pH by dosing acids / acid salts**

   In order to maintain solids in solution, water that is high in alkalinity and hardness may require the feed of sulphuric acid or an acid salt in addition to blowdown. This brings down pH of the circulating water to an optimum value so as to keep the Langlier index of the water to a desired level. However, acid feed requires careful handling and control and should only be used where the blowdownrate would otherwise be excessive. Excess sulphuric acid dose may also add sulphate ion to the water, which is required to be kept in limit (generally less than 600 ppm) against sulphate attack to the RCC structure. Improper mixing and uncontrolled dosing of acid may create serious problem of corrosion in steel structure and damage to RCC cooling tower. Polyphosphates like sodium hexa meta phosphate (calgon) with acid or without acid are of some value for scale control, but must be applied cautiously, because hydrolysis of the polyphosphate results in the formation of orthophosphate ions, which may result into calcium phosphate deposits. Hence, polyphosphates are now used primarily for corrosion control.

Since, some degree of supersaturation can usually occur without scale deposition, it is better to ensure that the water has a small positive index at the highest temperature in the system. A value of (+) 0.2 to 0.4 has been suggested, rather than take risk for corrosion.

3. **By softening of the make-up water or small% of circulating water:**

Under TPS practice, usually a base exchanger unit using strong acid cation exchange resin in sodium form are in convention, to reduce calcium and magnesium ions, which are prime factors for scaling. However, this process does not decrease the alkalinity of the water, and thus combination of low hardness, high alkalinity, high pH make the water more aggressive to corrosion particularly to copper alloys and notably to mild steel alloy and also when chloride and hardness imbalances are there. It has been observed for a surface water to lead considerable corrosion to mild scale material when carbonate hardness (as $CaCO_3$) are less than three times of chloride (as Cl). Additionally, acid feed may also be necessitated to control high pH after softening.

Removal of calcium sulphates, silica, fluorides and phosphates, iron, manganese, etc. are also required to be removed to an acceptable level, 'Precipitation softening' in place of base exchanger is employed, which is rarely a practice in TPS, due to the reasons of non-requirement as per raw water quality available to sight or due to the technical difficulties of facing high pH and then bringing to optimum level by dosing acids. For the purpose, a combination of lime and soda ash, along with coagulant and flocculant chemicals is added to raw water either in cold or hot, to promote a precipitation reaction, which allows softening to take place. Uncontrolled operation may likely lead to carryover of lime particles to foul the cooling system.

4. **By using proprietary cooling water chemicals:**

In view of water scarcity at site, since it may not be always possible to waste cooling water by blowdown, some proprietary chemicals are in use to deal with the scaling & corrosion problem, even when concentration of dissolved salts are on higher side, i.e. more COC can be allowed by way of using some chemicals including the proprietary ones, which can be elaborated as below; COC can further be increased by employing titanium / stainless steel metallurgy in the heat exchanger of the cooling circuit.

Leading Scale Control Agents

Agents	Solubilisation	Dispersant	Crystal Modification
Lignin/tannin	--	yes	Yes
Starch/alginates	--	yes	Yes
Acids	yes	--	--
Polyphosphates	yes	--	--
Phosphonates	yes	--	--
Phosphorous esters	yes	--	--
Low mol wt polymers (Poly acrylates)	yes	yes	yes

The chemicals acting as scale control agents on basis of solubilisation technique are mainly acids, polyphosphates and phosphonates, as generally used under TPS practice.

Acids: Sulphuric acid is most commonly used to bring down the observed pH of the cooling water, so as to enable Langlier or other index towards neutral regime against scaling or corrosion. However, care has to be taken while dosing H_2SO_4, that the sulphate content (SO_4) should not exceed 600 ppm of the cooling circuit water to avoid sulphate attack of RCC construction.

Polyphosphate: The aim of dosing is to distort the crystal lattice of $CaCO_3$ & slow down crystallisation process giving soft sludge. This is mostly dosed in the range 2 to 5 ppm.

Organo phosphonate/organic polymer: These are dosed in the range of 2 to 5 ppm. The function of these are to keep calcium salt in solution even at high pH, high calcium concentration & severe scaling conditions and facilitates increase of COC organo phosphonates can be taken as substitute for polyphosphate, as they display better stability with greater effect and are found of two types:

a) **Amino Methelene Phosphonic Acid (AMP):**

This is found unstable in presence of chlorine and becomes more aggressive after destabilisation. It not only completely looses its scale control properties but aggravates scale formation. AMP degradation product is orthophosphate, which reacts with either calcium to form calcium phosphate scale or with soluble iron & corrosion products to form iron phosphate deposits.

b) **Hydroxy Ethylidine Diphosphonic Acid (HEDP):**

This is found sufficiently stable at prevailing level of chlorine in the cooling water and is widely used in the cooling circuit, but will degrade in presence of iron to contribute deposits by forming iron phosphate sludge or scale.

Both phosphonates, however, can be taken as chelating agent and usually attack steel or copper alloy. When phosphonate is low or there is absence of Ca/Mg hardness, corrosion attack on steel surface is must.

Active phosphonate, however, 3 - 5 ppm can take care of psi of even 4.0. Not only can they cause premature failure of copper alloys, but also their action can result in copper transport and consequent plating out on steel surfaces. Hence, copper corrosion inhibitor should always be used when phosphonates & phosphate esters are used for scale control particulary in presence of cooper alloys.

Take care of danger of calcium phosphate precipitation either by chlorine or corrosion products in the cooling system, organic polymer are used as dispersant to keep away deposition of calcium phosphate on the cooling surface. $CaCO_3$ can easily be controlled with low dosage of phosphonate & polymer. A substantial improvement is obtained with the use of zinc also, which proves effective substitute for the removed calcium. Synthetic organic polyelectrolyte as dispersants or stabilisers is also on use towards this. Though $CaSO_4$ does not generally become a concern in most of the cooling water, polyacrylate, penta phosphonate & tetra phosphonate can be fed to control $CaSO_4$ scale, if required, 3 to 5 ppm polyacrylate (low molecular weight of 1000 to 10,000) in the cooling water is enough to keep calcium scale from forming even under severe scaling condition. These are good dispersant for suspended material including $CaCO_3$ & $CaSO_4$. These take care of both carbonate & sulphate scale at high temperatures (over 250 ^{0}F). These are not degraded easily and are stable and soluble at high & low pH levels. This can work up to a psi of 4.5. An active concentration of 1 to 2 ppm increases $CaSO_4$ solubility by a factor of 20. Disadvantage of polyarylates is that they react with strong cationic materials such as biocides and make-up water flocculants, which remove them from solution and reduce their scale control capability. High dosage of this, greater than 100 ppm may precipitate calcium polyacrylate.

Crystal modifier chemicals:

Scaling & depositing of crystalline chemical compound in cooling system can also be controlled in two ways as below:

a) An impurity to be introduced into the crystals as they block further growth or introduces strain into the crystalline structure.

b) Ions that can be added are adsorbed at the surface of the crystal, slowing or interfering with its further growth.

The principle, therefore, makes larger particles to grow at the expense of the smaller, making cooling water unsaturated with respect to small particles, but supersaturated with respect to larger crystals - the solubility of particles with radii larger than 10 micron, remaining virtually constant. The suitable concentration of the crystallisation inhibitor thus successfully prevents precipitation depending on temperature, composition of salt and its degree of supersaturation. Thus, though the crystal modifier allows the scale to form, it distorts the crystal structure changing the scale to a non-adhesive sludge.

Two types of crystal modifier chemicals are mostly available (1) Polymaleic acids (2) Sulphonated polystyrenes. Both classes are compatible with chlorine but can be deactivated with strong cationic chemicals. Recommended dose is 0.5 to 2 ppm. The treated cooling water should appear turbid, indicating crystal modifiers are taking effect.

It has to be noted that none of the scaling indices used for the solubilising methods are applicable in crystal modification (when crystal modifiers are used) in spite of the fact scale is constantly being formed. But a major advantage with crystal modifier is that it permits higher COC. Lignin sulphonates, starch, alginates & tannin as used earlier (40 - 100ppm) were found thermally or biologically unstable. They have chlorine demand & react with iron to form insoluble deposits.

Corrosion

The general corrosion reaction of mild steel with cooling water proceeds like this:

$Fe \rightarrow Fe^{2+} + 2e^-$ (anodic)

$2e^- + 2H^+ \rightarrow H_2$ (cathodic)

$Fe + 2OH^- \rightarrow Fe(OH)_2$

The corrosion rate in a cooling water circuit is evaluated generally on the following lines:

For mild steel	For Cu / brass	For S.S.
Good <3 MPY *	< 0.3 MPY	< 0.1 MPY
Accept 3 - 5 MPY	Up to 0.5 MPY	---
Correct 5 - 7 MPY	---	---
Bad > 7 MPY	---	---
	1 *MPY = 1 /1000"/yr = 0.0253 mm/yr	

At first sight, in cooling water, corrosiveness is generally a characteristic of water with high CO_2 in conjunction with a low pH value and low alkalinity. With the advent of Langlier & Ryznar index, etc. a Langlier index of zero was suggested to have a cooling water to be free from scaling and corrosion. But as already discussed, since some degree of supersaturation can usually occur without scale deposition, it is better to ensure that the water has a small positive index at the highest temperature in the system. A value of about (+) 0.2 to 0.4 has been suggested, rather than risking for corrosion. However, this is not the only cause of corrosion in a cooling water system. Dissolved gases like O_2, H_2S, NH_3, Cl_2, etc. suspended solids, water velocity, presence of bimetals in system and micro/macrobial growth, presence of high chloride & sulphate and cooling water chemicals as used can also lead to corrosion sometimes in aggravating form too.

The following are the types of corrosion mostly encountered in a cooling water system:

GENERAL	LOCALISED/ PITTING	GALVANIC	FRETTING
Aqueous conditions uniform distribution of corrosion product can cause fouling also	Formation of concentration cells in crevices/dead areas underlying deposits/ microbial growth etc. cause metal to perforate in very short time	Two different metals in contact. More active metal corrodes rapidly protecting less active metal. Large cathode & small anode is undesirable	Two dissimilar metals tightly fitted but slight movement possible. Usually the harder of the two suffer fretting

The corrosion is further aggravated due to:

Presence of anodic and cathodic areas on a single metal caused by 'differential aeration', by which the areas where dissolved oxygen concentration is low, will become anodic and will corrode. The metal in contact with liquid of higher oxygen concentration will become cathodic & protected from corrosion. Rapid failures of rivettes are on this ground.

The removal of oxygen from solutions in which metal is corroding, means that the oxygen absorption reaction becomes impossible, and the only alternative reaction is the reduction of hydrogen ions to hydrogen molecule, i.e. $H^+ + e^- \rightarrow H$; and $H + H \rightarrow H_2$. In neutral & alkaline solutions, the number of hydrogen ions is small, therefore the cathodic process will take place at a slow rate, and the anodic process will be equally slow.

$$O_2 + 2H_2O + 4e- \rightarrow 2OH^-$$

Dissolved oxygen promotes general corrosion, and the rate of its arrival at the surface is more important. More water flow brings oxygen faster to the metal surface to increase corrosion & carrying away corrosion products to promote fresh corrosion at the surface and cause under deposit corrosion in low flow region. In aerated fresh water system, admirality brass UNSC 44300 is limited to water below pH 7.2 and a process inlet temperature not greater than 350 ^{0}F for cooling service.

Total dissolved salts & composition of water and the area of the two bimetals determine the extent of galvanic corrosion. Higher the conductivity of water, wider is the area of attack. It becomes progressively more difficult & expensive to inhibit corrosion as the sp. conductivity approaches & exceeds 10, 000 micromhos/cm.

Chloride at pH above 8.2 if not supported by due hardness cause dezincification and accelerates general corrosion. Both high chloride and sulphate are corrosive to mild steel surface, but chloride in lesser amount also could be dangerous for stainless steel at prevailing temperature of the cooling water.

Presence of dissolved oxygen, if it is equal to or below about 0.1 indicates general freedom from corrosion, whereas increasingly higher ratios generally indicate more aggressive water. The chloride ion content of water limits the use of stainless steel 304 SS & 316 SS. The maximum limit is respectively 200 & 500 ppm for water at pH 6 & temperature less than 140 ^{0}F.

Residual chlorine, which affects oxidation - reduction potential, also nitiates corrosion. This forms both HCl & HOCl acids in the cooling water. Delignification of timber is often severe when high concentration of chlorine & alkalinity occur together. Cascading water aggravates the deterioration by washing away the surface fibres. The loose fibres settle down in cooling pipes when under shutdown & in due course choke the cooling water exchanger.

Every 3^{0}C rise in temperature doubles the rate of corrosion. Similarly high velocity water causes more erosion-corrosion. A velocity of 2 m/sec is optimum for cooling water pipes except at some points in pre-screening system culverts, where it can be reduced to 0.5 to 1 m/sec. The velocity is reduced further, if total solids or gas concentration are high, otherwise more impingement attack (erosion-corrosion) is feared.

Fouling or deposits including of biological origin generally initiate localised or pitting corrosion. When algae become brown, means they are killed, but the same, if not removed from the system, provides food for remaining bacteria. Most of the bacteria are (negatively charged & hence are attracted towards metal anodic in nature. The desulfovibrio (SRB) or sulphate reducing bacteria (SRB), which reacts with sulphate content of water in absence of oxygen (anaerobic condition) are giving H_2S to give ferrous sulphate with steel, which in turn becomes pitted & cast iron becomes graphitised. These are decomposition product of other bacteria or algae, area of growth being under deposit. The corrosion pattern is smooth, dish shaped concentric rings. The bacteria are extremely aggressive even in presence of good corrosion inhibitors since they are covered with slimes & envelope of reducing environment of H_2S. Iron bacteria absorb oxygen (aerobic) and oxidises iron. These remove water soluble ferrous iron salts to deposit as ferric iron. Nitrifying bacteria reacts with ammonia to produce HNO_3. Thiobacillus converts soluble sulphur compounds including H_2S to H_2SO_4. These acids as generated by bacteriason the surface of metal cause biochemical corrosion of ferrous & copper alloy tubes. Copper-nickel alloy should not be used in water containing greater than 0.7 ppm sulphides. Aluminium-bronze alloys have some what higher tolerance for H_2S. Aluminium-bronze or Cu-Ni alloys are often used when the process inlet temperature is 350 to 450 ^{0}F. FRP material has less degree of slimes than Fe, Cu, etc.

Some scale & deposits control chemicals including the properietory type if not fed properly or with prior assessment can also be responsible to aggravate corrosion.

Metals in salt and fresh water corrode at similar rates. However, salt water produces higher penetration rates (25 mils/yr in oxygeneted sea water) compared to 10 mils/yr in fresh water. The only practical way to reduce corrosion so that bare carbon steel can be used in salt water is deaeration and addition of an oxygen scavenger to keep oxygen below 20 ppb (0.02 ppm). The penetration rate of carbon steel in deaerated salt water is 3 mils/yr at pH 6.5 or greater, but corrosion rates increase rapidly below pH 6.5. Scaling begins above pH 7.0. Titanium, which is non-toxic, might be expected to be affected by the growth of organic slimes/films, but copper has some biocidal activity.

Corrosion Control

The following are the ways to control corrosion in a cooling water system, which vary from site to site in view of prevailing water quality and the metallurgy employed in the cooling circuit:

1. After an elaborate survey, it is found that Cu-Ni alloy (70/30) is best for sea water & admiralty brass dominate fresh water. However, nowadays, titanium tubes for sea water and stainless tubes for fresh water are also used.
2. Limiting pH of the cooling circuit water by suitable dose of acid/alkali, to get slightly positive LI (Langlier index), i.e. having little depositing tendency. Uncontrolled addition of acids or chlorine also gives higher risk of corrosion.
3. Limiting COC (cycle of concentration) to minimise the corrosive potential like chloride, sulphate, degree of alkalinity/acidity, etc. The COC may however be increased by dosing antiscalant/anticorrosive agents or employing titanium tubes, which can withstand corrosion. At pH 6 & temp. 140 ^{0}F, Cl content needs to be limited to 200 ppm for 304 SS and 500 ppm for 316 SS.

   If the sulphide content of the cooling water exceeds 0.7 ppm, severe corrosion of Cu-Ni alloy is anticipated. Al-bronze has higher tolerance for H S (temp. 350 - 450 ^{0}F). For admiralty brass UNSC 44300 pH $<$ 7.2 & temp. not greater than 350 ^{0}F.
4. Aerate gaseous impurities like oxygen, carbon dioxide, ammonia, hydrogen sulphide, chlorine, etc. The penetration rate on carbon steel for corrosion in deaerated water at pH 6.5 is 3 mils/yr. Below pH of 6.5 there is rapid corrosion and at pH $>$ 7 there is scaling. Hence, it is beneficial to get oxygen removed even by using scavenger. Application of electrostatic, magnetic or ultrasonic fields to modify 'settling properties' of particles, viz., turbomag system, etc. should be undertaken very carefully, since sludge created by these act as a very good breeding ground for microorganism, which in turn, impair heat transfer and increase corrosion.

5. Limiting chloride-hardness ratio to avoid complete soft water in the cooling system, which is highly corrosive, leading to dezincification from galvanised iron or steel, if the cooling water pH is above 8.2. Maintaining the following ratio is generally helpful:

Chloride (Cl ppm)	Carbonare hardness $CaCO_3$ ppm	Ratio Cl/hardness
10	10	1.0
15	15	1.0
20	35	0.57
30	90	0.3
40	120	0.3
60	150	0.4
100	180	0.56

Thus, if there is more chloride content in the water, simply raising pH is meaningless unless hardness is also present.

6. Coating metallic surface, which is in contact with cooling water, with epoxy paint, coal tar, asphalt, lead, etc. It has been noted that water with fairly high carbonate hardness due to $Ca(HCO_3)$ are seldom appreciably plumbo solvent, even if they have slightly negative Langlier index. Reason for this is the low solubility of lead carbonate whose solubility product is 3.3 x 10 - 14 compared with 1.0×10^{-8} for calcium carbonate. Ferrous sulphate dosing in the range of 1 ppm dose are also known to create few mils thick film of ferrous ion to prevent corrosion of the metallic surface.

7. Cathodic protection by using

   a. Sacrificial anode like Zn, Al, Mg etc. at base end of standard galvanic series or by employing.

   b. Impressed current making 3rd electrode (viz., platinised titanium to make anode for whole system. For practical purposes, separate 'galvanic series' as below are to be drawn for particular environment, e.g. Al & Ni are less base in sea water than indicated by the e. m. f series (standard electrode potential). Titanium, which is at the base end of the emf series, behaves like a noble metal under sea water conditions:

Noble end	Base end
Standard	Au, Pt, Hg, Ag, Cu, H, Pb, Sn, Ni, Co, Fe, Zn, Mn, Ti, Al, Mg, Na, Ca, K
galvanic series	-----------------Active metal-----------------
do under sea water	Ti, Monel, Passive S. S, Ag, Ni, Cupronickel, Al bronze, Cu, Alphabrass Al brass, Munz metal, Sn. Pb, Active S. S, cast iron, mild steel, Al, Zn, Mg, etc.

As per the following reaction,

$$Fe\ (Metal) + 2\ H^{+}\ (Ion) + Fe^{2+}\ (Ion) + 2\ H\ (atoms)$$

The positive ions can go into solution in an electrically neutral system, only if an equivalent number of positive ions of some other element are plated out. In case of pure water in contact with metal iron, hydrogen ions are plated out at the interface. The liberated hydrogen forms a thin insulating film on the metal, which prevents further corrosion. Thereafter, the corrosion reaction proceeds at a rate that is a function of the rate of removal of the layer of atomic hydrogen. Thus, any method of maintaining the protective hydrogen film on steel cathode will reduce corrosion, either in the form of sacrificial anode or a corrosion resistant anode.

In general, galvanic corrosion can be prevented by adopting the following:

a. Select materials close in galvanic series
b. Avoid unavoidable areas ratio
c. Insulate dissimilar metals
d. Apply coatings preferably on more noble metal
e. Reduce the aggressiveness of the medium
f. Design for ready replacement of active metal

It is, however, to be noted that using anodic inhibitor or cathodic alone, entire corrosion potential is shifted to unprotected corrosion sites causing severe pitting.

8. Dosing corrosion inhibitor chemicals including proprietary chemicals, which generally give non-conducting barrier in form of thin adherent layer/scale on the metallic surface in contact of the cooling water. There are three types of inhibitors as below:

ANODIC	CATHODIC	ORGANIC BASED
Chromate Nitrites	Polyphosphate Copper sulphate Zinc	Polymer Lignin Phosphonate
Ortho silicates Ortho phosphate	Molybdate Poly silicates Ferrous sulphate	*Copper corrosion inhibitor like as below:

***Copper Corrosion Inhibitor** as generally mixed with other above inhibitors:

Mercapto Benzo Thiazole (MBT),	**Benzo Triazole (BZT) &**	**Tolytriazole (TT)**
General dose 10 - ppm	1 ppm	1 ppm

All above are excellent for copper, but are also fair for steel and prove good for aluminium with working pH 6.0 to 10. However, the inhibitors are reduced in presence H_2S, SO_2 and hydrocarbons.

Combination of both anodic & cathodic inhibitors gives 'synergistic effect'. Using anodic inhibitor alone, entire corrosion potential is shifted to unprotected corrosion sites causing severe pitting. Also lesser amount of cathodic inhibitor gives general attack of corrosion. Organic based inhibitor treatment has been found technically more acceptable, even when it is less effective than inorganic one. The best one among inorganic had been the chromate based inhibitor, but its use has been banned totally from pollution control angle. Similarly, one has to be very cautious before using any heavy metal based inhibitor. Negative aspects of other inorganic based inhibitors can be seen as below:

Nitrite	1st anodic inhibitor used with polyphosphate (@500 ppm at pH 7.5 avoids pitting attack), found excellent for closed cooling system, but unsuited for open recirculating, in view of their oxidation product providing a nutrient to bio organism. At this concentration, they are toxic to animal life.
Copper	Gives plating on steel
Zinc	Gives fouling at pH > 8.0 & dose > 1.5 ppm. However, is good for steel, but neutral to Cu & Al. Activity reduces in presence SO_2 & hydrocarbon. pH limited to 6.5 - 7.5 & TDS < 5000ppm Zn is nutrient to microorganism at concentration < 1 ppm, but above 3 ppm is toxic, which cannot be dosed in view of its solubility which is not greater than 1.5 ppm at high pH.
Polyphosphate	Gives sludge. Excellent for steel, but attacks Cu & Al. Activity reduces in presence of H_2S, SO_2 & hydrocarbons. Working pH limited to 5.5 - 7.5 and calcium concentration 100 - 600 ppm.
Ortho Phosphate	Reacts with calcium and iron to give sludge, which becomes a potential scale former on pH fluctuations Have reduced protection in low velocity water (< 1 ft/sec), but good for 3 - 10 ft/sec. Do not protect Cu & Al. Ortho phosphate is also good nutrient for algae.
Poly silicate	Adverse effect is there if water contains SiO_2 > 10 ppm. For steel, Cu & Al, it is excellent, but reduces in presence of H_2S, SO_2 & hydrocarbon; pH limit 7.5 - 10 & TDS up to 5000 ppm.
Molybdate	There is good success at pH > 8.0 and < 10, but adverse effect is noted in water having TDS > 5000 ppm. Good for steel, fair for Cu & Al. Activity reduces in presence of SO_2 & hydrocarbon.

Hence, acceptable formulation of any corrosion inhibitor should be made as per water quality. After checking the sensitivity of the formulation, it is necessary to add 2 to 4 times of normal dose for first few days after suitably cleaning the surface.

No corrosion inhibitors are however found satisfactory above a TDS of 20,000 ppm & calcium > 1500 ppm, and sea water crosses the limit in general. Typical *Non- chromate corrosion inhibitors* can be seen as below:

INHIBITORS	RECOMMENDED DOSING RATE
Polyphosphate	10 - 30 ppm as PO_4
Ortho / Polyphosphate (1:1)	10 - 30 ppm as Total PO_4
Polyphosphate / Zn	10 - 30 ppm as PO_4 / 1 - 5 ppm as Zn
*Zinc / Organic phosphonate	3 - 5 ppm as Zn / 3-5 ppm as phosphonate
Zinc / tannin / lignin	3 - 5 ppm as Zn / 50-100ppm tannin + lignin
Polysilicate	10 - 15 ppm as SiO_2
Molybdate / Phosphonate	5 - 20 ppm MoO_4 / 3 - 5 ppm phosphonate
Poly silicate / Molybdate	10 - 20 ppm SiO_2 / 1 - 3 ppm MoO_4
Phosphonate / Polysilicate	5-20 ppm phosphonate / 10 - 20 ppm acrylate

*Serves deposits inhibitor also, but any zinc formulations are generally not advisable to be used in closed circuit of cooling system in view of its fouling characteristics at concentrarion > 0.5 ppm.

Many of polyeloctrolytes in the market which have one or the other advantage are also made making use of 1, 2 phosphonobutane, 1, 2, 4 tricarbonic acid (PBTC), which creates film inhibition (and not boundary layer inhibition) to corrosion; i.e. layers of finite thickness are formed on the metal, which stops corrosion process (in contrast, a boundary layer inhibition by some other chemicals is due to mono molecular deposit on the surface). The layers are basically built up from PBTC in the form of Ca/Fe (ii) compounds. At a high pH value & high buffer capacity due to stabilised carbonate hardness, a highly effective corrosion inhibiting film can be created through effective organic stabilisers in organic formulations, free from heavy metals. Further, PBTC has the advantage of chlorine stability.

Thus, based on the deposits/scaling and corrosion problems, as faced by different utilities, on their cooling water system, efforts should be made to have the following circulating water quality, if possible, for trouble free operation of the cooling system. For an increased concentrations of the ingredients in the water or to have increased cycle of concentration (in view of scarcity of water), suitable dose of antiscalant/dispersant and corrosion inhibitor are found recommendable. With dosing of antiscalant/dispersant and corrosion inhibitor, hardness, LSI, silica, chlorine and TDS can go significantly high along with water saving.

pH to	suit L. I. around (+) 0.2
Turbidity	not greater than 50 NTU
Total hardness as $CaCO_3$	not greater than 250 ppm
Temporary hardness as $CaCO_3$ chloride as Cl ppm:	not greater than 200 ppm
Temporary hardness ppm do	approx. 1:3
Iron as Fe + manganese as Mn	not greater than 1.0 ppm
Total dissolved salts	not greater than 300 ppm
Organic matter ($KMnO_4$ No) mg of O2 absorbed; 37 ^{0}C; 4 hours	not greater than 2 ppm
BOD	not greater than 4 ppm
Carbon dioxide as CO_2	not greater than 25 ppm
Chloride as Cl	not greater than 200 ppm
Sulphate as SO_4	not greater than 70 ppm
Total silica as SiO_2	not greater than 75 ppm
Bacteria	TBC < 1 lac/ml
	SRB < 100 org/100 ml

However, during the operation cycle of a cooling system if some upsets with respect to the following are noted, corrective measures as noted below need to be taken:

System Upsets & Corrective Measures

1. **High pH of circulating water:** If noted, it is necessary to investigate for corrosion. There could be improper or no acid feed; increase in make-up water pH; more COC; ingress of alkaline contaminants, etc. pH is reqired to be controlled below 8.5, but never to exceed 9.0. In the event, though corrosion rate decreases, but both scaling & microbial growth increases. Hence, increase in concentration of scalant/dispersant dose & also shock dose of biocide may be needed if contaminant ingress is there.

2. **Low pH:** If noted, means excess feed of acid; ingress of acidic contaminants like flue gas, carbon dioxide, sulphur dioxide, hydrogen sulphide etc; development of sulphate reducing & nitrifying bacteria.

   In the event of problem occurring too frequently, operate the system on lower COC If the pH falls below 6.0, repassivate the system. At pH 6.0-7.0, increase the concentration of corrosion inhibitor by 25-30%. Dose soda ash/caustic gradually to increase pH.

3. **Low inhibitor concentration:** There could be improper feed of chemicals; degraded chemicals, excess blowdown & fresh make-up; presence of absorbent matter like fly ash & carbon absorbing organo phosphonate.

   In sufficient centration cause more scaling, fouling and corrosion. If situation persists for longer time, acid cleaning & passivation of system may be needed.

4. **Ingress of oil:** Source is needed to be located by determining oil & grease, COD, TOC or $KMnO_4$ of water & removed. Apart from nutrient for microorganism, oil acts as binding media for suspended solids. Hence, suitable surfactant dose is necessary.

Sea Water Cooling

Many thermal power stations have now to depend on sea water for their condenser tube cooling in view of huge water requirement for this and fresh water is unavailable for the purpose. This is in spite of the fact that for using sea water, both condenser surface area and CW flow rate have to be increased by 5-8% and 6% respectively in comparison to the river water, if used. For some power stations, even sea water is also available in limited quantity, and sea water cooling with cooling tower is also now offing. Hence, attention is now being paid on suitable chemical treatment of sea water to avoid fouling/ scaling and corrosion in its cooling systems & equipments, although no chemicals including properietary ones have so far come up specifically for sea water cooling.

In once through cooling system by sea water, scale formation is not normally a major problem, nor bacterial slimes due to microorganisms are feared in presence of high amount of NaCl (3 to 4%), but high TDS including very high chloride & sulphate etc content make sea water highly corrosive, for which specific metallurgy including cupro nickel (70/30) & titanium have to be envisaged for the equipments handling sea water. In recirculating type of cooling system, the cooling tower has also to be made with specific infrastructure including use of very good sulphate resistance cement. Under this cooling system, since cycle of concentration is generally opted from 1.5 to 2.0, all problems of fouling, scaling & corrosion can be encountered. Still micro fouling is not a major problem, but macro fouling is abundant. The sea water, which is having some phosphate content also above 5 ppm can be a good source of calcium phosphate scaling. If sea water is from estuarine or from a source of industial pollutants, etc. again all sorts of problems can be encountered. In general, therefore, macrofouling is the main concern to be addressed, during operation of a cooling system in sea water, with scaling and microorganism fouling in specific cases only. Erosion & corrosion problems are duly taken care of by providing suitable metallutgy during design stage itself.

It may however, be mentioned in this connection that natural draft cooling tower for large duties should be prefered to avoid corrosion of rotating equipments and other equipments in the surrounding areas due to drift from tower on forced draft. The condenser surface area and CW flow requirement have however to be increased due to increased salinity of water causing lower vapour pressure resulting to reduction on heat transfer characteristics by 2 to 3%. This then also requires an increase in air flow rate & consequently fan kw rating of a forced draft tower in comparison to same working on usual river water.

Macro Fouling

Although, there are atleast 2000 different species of organism causing marine fouling & nuisance, which can be categorised in ten groups as below,

1. Algae
2. Bryozoans
3. Barnacles
4. Hydroids
5. Amphipods
6. Sea Anemones
7. Mollusks
8. Sponges
9. Tubeworms
10. Tunicates

The main fouling organisms in sea water cooling systems of TPS, of concern, apart from algae, worms, snails, crabs & shell fishes etc are found as:

1. Bivalve mollusks/Mussels (mytilus adulis as typical)
2. Crustaceans/Barnacles (cirripedia as typical)

In cooling system, though large & adult mussels, barnacles, etc are strained away by different type of screens at different stages, these will develop from spat (larvae) to shell forming stage and settlement in different areas of the cooling system, in some 70 hours or may be sooner in warmer water. The attachment however starts with the roughening of the metal surfaces by erosion/corrosion (bacterial counts often reach 104 counts/ml). These then provide a suitable substance for attachment of larger marine organism such as diatoms and in the final stage, the larvae of mussels, barnacles, worms, etc.

The mussels may grow up three times faster in darkness; warmth will also increase the growth rate. The female may lay up to 25 millions eggs during the life time and after 24 hours, the eggs become free swimming spats. 70 hours after the egg is laid, the spats (0.4 - 1 mm in size) will attach to the walls of culverts & mains within a few minutes by means of tough sticky secretion (byssus) which hardens in the water to a hard thin strong elastic threads. *Fauna* present influence along with turbulence & low velocity of water clinging of the larvae to the conduit walls & commence adult life. The shell is developed by this time. Spawning takes place between mid April & November around North Sea. *Mytilus edulis* is a major fouling mussels on the east coast, typical for all fouling mussels. Hundreds of tons of mussels accumulate in an intake system of a power plant in cooling water culvert up to 3000/m^2 of these creatures have been counted. This has a wedge shaped shell & is rarely more than 2 inches long. The shell is smooth. When they die, mussels detatch from the walls and the shells are carried by water to condensers, clogging the tubes. Apart from barnacles, hydrozoa also attach to the concrete walls & water mains and feed on plankton incoming with fresh water.

Barnacle is an animal, capable of colonising surfaces on which other marine animals cannot exist and it lives in deep water as well. They are grouped with such animals as crabs, lobsters and shrimp.

A barnacle is described as being a little 'shrimp' in a shell, standing on its head & kicking its feet. Appendeges coming out of the shell opening is called 'cirri', which brings in food & water to the mouth. The barnacle feeds on small plants & animals in the plankton. The most common fouling barnacles on each of the coasts is genus 'balanus' with several different species of which 'acorn barnacles' and goose neck barnacles are common.

A corn barnacles attach the shell directly to a rock or other substrate. The goose neck is found on pieces of rope or wood, and is more readily removed. These acorn barnacles are among the most serious fouling problems on ship bottoms, which reduce the speed even up to 30% with badly fouled ships. These ships then are able to foul other coastal areas wherever they are langered.

Eggs & larvae of these acorn barnacles are retained within the shell of the parent and are discharged as free swimming animals called 'Nauplii', which is a one eyed, one shelled, microscopic animal with 3 pairs of legs, and it swims with the plankton. After going through 6 or 7 Naupliis stages, it becomes a 'Cypris' with 3 eyes, 2 shells, & 6 pairs of legs. These Cypris larvae then can delay metaphorphosis for as long as two weeks, should it not find a suitable surface. These attach to a suitable substratum, at first temporarily with the secretion of discs located on the first antennae and then permanently, with the antennal cement glands. When the larvae makes contact with an individual of its own species, viz., barnacles already present on this surface, the cypris larvae will space out a certain distance from the other individual before it settles. If the larvae make contact with an individual of a different species, it will settle on the shell of this individual. This behaviour helps to prevent crowding of the individuals of the same species and provides completion of other species.

The young barnacle is about 3 mm in diametre at the end of a month. The maximum basal diametre of the adult is 22 mm of which 7 to 12 mm is reached by end of 1st year and 14 to 17 mm at the end of the third year. The life span is 8 to 10 years. Most barnacles that survive the heavy mortality in the period immediately following settling, probably live for 1 to 10 years.

Live mussels & barnacles along with marine algae are found generally in intake channel, culverts & the cooling tower basins & body of intake pumps, but rarely in the heat exchanger tubes except in the region of low velocity of cooling water. The dead species along with their shells have however been found generally choking the cooling tubes resulting to adverse effect on electricity generation.

Marine algae make-up the greater part of vegetation of ocean and can be seen as common sea weeds, free floating or attached to the substratum by the lower most cell or the hold fast. The free floating marine algae are generally unicellular or colonial form and are spoken of collectively as 'phytoplanktons'. These plants constitute the basic food supply on which the fish & other marine animals feed. Apart from this type of algae, thousands of species of other marine algae could also be found, commonly known as brown & red macro algae (phaeophyta & rhodophyta). These species can settle with a little base on the surface developing a vertical frond, so that to hide the light to the lower organism. They can live better at low levels of light and can develop croustous thalli. The fouling layer reaches the maximum development with adhesion of marine animals such as crustacean.

Thus, biofouling in sea water cooling system starts with algae and complicates in general with crustacea & bivalva. The most important crustaceous fouling species is balanus (barnacle) that has planctonic larvae that can produce at the surface a strong extra cellular polymeric substance (EPS) called cement & adheres strongly to many materials. These EPS can further be produced by algae, fungi, diatoms, protozoa, bacteria, etc. also, responsible to create biofilm to favour further adhesion to cause a complex mixture of suspended solids, corrosion products, microbial masses, etc., and these biofilm is increasingly difficult to remove the older it is. EPS on deposition & redeposition of pseudomonas aeruginosa to surfaces could be detected as bacterial footprints of uronic acid.

Biofilms are more resistant to biocides than suspended bacteria as microorganisms in the central layers are more protected by means of their EPS. Biocides have to break the physical stability of the biofilm matrix allowing the shear forces to remove the biofouling layer & preventing the transition from micro to macrofouling. Slime layer on condenser tubes, if found, would potentially enhance early 'macrofouling' & significantly reduce station heat rate. The stability & adhesiveness of biofilm matrix is therefore the crucial factor, if we want to keep the extent of biofouling deposit accumulations as low as required by majority of industrial processes.

Biofouling Control Measures

By Strainers: Any control programme can readily control immature forms, but adults are virtually impossible to remove. Further dead algae which become brown provide food for living bacteria. To remove therefore clogging by debris & other macrofouling, it is necessary that first the 'bar screens' with bar distance of 100 - 150 mm is installed and then supported with fine screens with bar distance around 30 mm and finally 'travelling water screen' with mesh widths of 2 to 4 mm, on the path of cooling water flow to the cooling system. Nutrient feed is supposed to be better, if T. W. screen size is greater than 5 mm and hence should always be less than 4 mm (1/8").

By Biocides: The efficacy of biocides depends on:

1. Kind of biocides
2. Mechanism of action
3. Concentration
4. Kinetics
5. Dosing methodology

For all the above foulants, intermittent chlorination of 0.2 ppm TRO (total residual oxidant) has been found sufficient to prevent biofouling with continuous biocide feed during the winter months to prevent slime formation on heat exchangers. Low level continuous chlorination, (0.05 to 0.10 ppm as TRO) almost completely eliminated all sessile fouling species except barnacles. Barnacles fouling increase with continuous chlorination vs no biocidal treatment at all. No advantage further on barnacle control along with general fouling on increasing TRO from 0.05 to 0.2 ppm have been noted. At concentration of 1.0 ppm of TRO, however, mussels have shown 100% mortality.

A 12 hour off feed has been found as effective, as continuous feed at 0.2 ppm TRO. Also, four 30 minute shocks at 5.0 ppm TRO is as effective as continuous feed at 0.5 ppm TRO. Hence intermittent shock chlorination is better than continuous chlorination. The continuous chlorine dose generally controls other fouling species with intermittent having lesser effect. On low flow or non-copper alloy system, barnacle population actually increases with either continuous or intermittent chlorination.

Chlorination is however not a complete solution due to unequal flow condition, dead spaces, turbulence in the culvert and fluctuating loads reducing cooling water velocities. Colonies of mussels/barnacles up to 200/m^2 may even be found. Continuous chlorination however causes the mussels/barnacles to produce less byssus. Chlorine will not be helpful where there are dead species.

At high pH, chloramines are just as effective as chlorine against mussels/barnacles/algae etc.1 - 2 ppm of combined chlorine residual in form of chloramines maintained for a larger contact time can control biological fouling. However, ammonia feed equipment or a dual application system must be installed. Sodium hypochlorite is also the choice for small system which provides the same kill rate at low dosage. But all these form trihalomethanes (THM) which are potential carcinogens. Using sodium hypochlorite has another problem of precipitating out insoluble magnesium hydroxides out of magnesium salts present in sea water, due to high pH content of the sodium hypochlorite, which in turn can deposit in the condenser tubes. However, for biofouling control & system availability could be better from NaOCl than gaseous chlorine. NaOCl: NaBr (6:1), a surfactant solution are better than NaOCl alone. However, NaBR is not required in case of sea water, which contains bromide.

Other biocide compounds, which do not form THMs and have effective control on macrobiological fouling are:

1. **Chlorine dioxide (clo_2):** It is more effective than chlorine. Produces 100% mortalities @ 0.5 - 0.6 ppm dose, but is expensive and must be

generated on site from sodium hypochlorite. This performs significantly better than NaOCl both on veligers & adult mussels/barnacles, along with depletion of their food sources (i.e. planktons) and creating unsuitable environment for their settling in view of disruption in swimming patterns and internal mussel movement.

Though ClO_2 has clear advantage of having strong oxidant, good oxidizing biocide over a broad pH range & have no THM formation and does not contribute to change in system pH, yet other negative points as below need to be addressed as per the site situations:

a) Environmental angle, the discharged water from the cooling system needs to be suitably dechlorinated, except when warmer water temperature & very low concentration (<.5 ppm) ClO_2 are in use. ClO_2 does not dissociate in water, but forms true solution of gas in water. Its actual demand in water & assessment of its residual concentration are uncertain, thus organic molecule & ammonia, the traditional chlorine demand material with which ClO_2 does not react, remain untreated in the discharged water.

b) ClO reacts aggressively with some of chemicals as used to prevent deposits, corrosion, etc. like TTA, BZT, MBT, lignin, organics, etc.

c) It has more oxidising power than chlorine, but at water pH above 8.5. Hence, it is difficult to assess the effect on sea water, which has generally a pH of 8.0 and less.

d) It is highly unstable. It cannot be stored, requires generation online with 2 - 3 numbers reactants & the efficiency of generation is highly dependent on concentration of reactants, pH, temp, reaction time & mixing efficiency. Still on maximum efficiency, the yield is only 75% & 25% of NaClO (purcursor) is unconverted, the dried form of which can spontaneously ignite.

e) It is 700 times more volatile than Cl means a toxic material has more loss from cooling tower system. An exposure can be fatal within 45 minutes with levels as low as 150 ppm and least detectable odour is at 14-17 ppm. It explodes at 14000 ppm level. But it can be generated safely, already demonstrated worldwide.

Hydrogen peroxide and ozone are other oxidising biocides which can be used in place of Cl_2 and ClO_2, but are not cost is very cost effective at 9 ppm dose and ozone at 0.5 ppm, but with 95% kill. Also on macrofouling these biocides have no proven data for its effectiveness. More than 5 ppm dose of H_2O is not recommended when water has to be used for drinking purpose.

Non-oxidising biocides:

Among these groups mostly are of properietory type chemicals including of quaternary ammonium compounds, polymers, molluscides, etc. and have been found effective on macro biological fouling too on low dosage and at moderate prices, but methods to kill veligers (larvael stage of mussels/barnacles) offer greater benefit than focusing on adult eradication. While the oxidising biocides toxicity can be sensed easily by mussels/barnacles veligers, the properietary non-oxidising chemicals' toxicity cannot be detected. They keep filtering the same and die. The chemical, however, in suitable concentration must be in contact with the veligers for a minimum of 8-12 hours.

By other choices of control: Among other choices of controlling macrofouling concerning with use of sea water in cooling system are with the following steps:

a) Raising temperature of cooling water cycle up to 40 ^{0}C several times per year, for a period of several hours. This step kills mussels & other boilogical life in part, but there may be inadmissible heat stresses.

b) Changing water quality (e.g. pH & salinity). Altering sea water salinity do not allow mussels/barnacles etc to develop critical size with respect to condenser tubes to clog them. Some even die of osmotic influence.

c) Raising the velocity of water up to 3 m/sec. This will not allow mussels/ barnacles to settle, but the constraint is load & design data. In view of this fluctuating load needs to be avoided.

d) Reduction of oxygen content in the circulating water, but care must be taken to avoid aggravating system corrosion

e) Reversing the flow of water, but mussels/barnacles already clinged is not removed.

f) Installation of online cleaning system like sponge rubber ball cleaning/ taprogge debris filter type etc.

g) Toxic painting of all parts in contact with cooling water reduces settling of mussels & other aquatic organisms, but it is difficult to repaint in the humid environment, when required on regular interval. After being immersed for six months, only silicone polymers proved effective, showing a weak attachment of mussels coverage of less than 20%.

h) Acoustics (e.g. ultrasonics), which are also supposed to kill some of the organisms.

i) Off-load cleaning of serious infestations solidly clogging the tubes by clams, barnacles, mussels by steel balls & guns, but expertise is needed to operate the guns that drive these clinging organisms out of the tubes.

j) Avoid fluctuating load to reduce water velocity to facilitate the settling of the organisms.

k) Titanium, which is non-toxic, might be expected to be affected by the growth of organic film/slimes. Copper has some biocidal activity.

l) Continue chlorination, due to which mussels will produce less byssus. However, chlorine will not help in dead spaces. Dead algae become brown and are food for living bacteria and other biomass.

Monitoring & Evaluation of Treatment Programme

For Corrosion: Suitable corrosion coupons have to be installed in return header line of the cooling water circuit, i.e. in the water coming out of the heat exchanger. Exposed area of the coupon needs to be 1300-2600 mm^2. This has to be then taken out of the header after 30 - 45 days. The coupon has to be of same material & mounted with accessories of plastic/phenol formaldehyde material.

For Scaling & Fouling: It is difficult to measure. Assessment needs to be made indirectly to heat transfer data & periodic inspection of exchanger when under shutdown. Scale & deposits, if any found, have to be analysed for the quantum and its nature to take corrective steps accordingly. However, for scaling 15-30 mg/cm^2/yr of deposition is expected. Any thing beyond this is a cause of concern.

For Microbiomass: Analysis of circulating cooling water before & after dosing of biocide needs to be made, with examination of cooling tower louvers & deck for algae & fungal growth and evaluate biocide effectiveness accordingly.

In the laboratory, a tentative evaluation of water for total bacterial count (TBC) and sulphate reducing bacteria (SRB) may be viewed by 'bacta slide' by dipping the same in the water under examination for a specified period under specific temperature as guided by the supplier, and the deposit on the slide compared with the density chart as supplied along with the slide.

The percent kill of greater than 80% indicates good biocidal effectiveness with TBC $< 5 \times 10^5$ & SRB $< 100/100$ ml.

For Macrobiomass: It is difficult to assess except examining the condenser tubes & cooling tower basin under shutdown or there is vacuum drop of condenser due to choking of the tubes. A biobox made of plastic or phenol formaldehyde may also be installed at the return line of the cooling water, in which few RCC plates are kept immersed in flowing water through the box. The plates are needed to be examined every month to see the population growth of the macrobiomass, if any.

Conventional biocides as used for microbio masses are still in use for sea water generated macro biomass too, but not with proven success so far. Experiments with various combinations are still in progress to deal with the problems of fouling by sea water cooling.

Chapter - D

WATER POLLUTION CHEMISTRY

(Describes possible impacts of waste water emanating from a TPS on the environment, and means to contain them. Also, highlights impacts of various elemental impurities as could be present in any other waste water)

Water Pollution Chemistry

For a thermal power station, the word 'water pollution' comes with the waste water generated by it. The treatment of waste water for discharge and for recycle are complementary approaches to waste water management i.e. conservation ethic can be tied to water pollution control.

The objective of pollution control should be that there is no impact of a TPS operation on the project area & surrounding area up to 30 km radius with respect to existing base line environmental conditions, and implement sound environmental safeguards accordingly or as suggested / directed by Pollution control authority / Environmental protection agency (EPA)

The chemical wastes as addressed generally by the pollution control agencies, consist of (a) acidity / alkalinity (pH), (b) suspended solids (c) oil & grease (d) PCBs (poly chlorinated biphenyls) (e) chlorine, copper iron, zinc, chromium, phosphorus and scale/deposits contol chemicals & corrosion inhibitors, etc. Toxic pollutants are also covered by the agency, with the specified list generally based on primary drinking water standards initially, for determining toxicity for pollutant elements, like arsenic, cadmium, chromium, copper, lead, mercury, nickel, selenium, zinc, barium, silver, etc. and then for safe guarding the other ecological features. However, it is generally agreed that the large majority of toxics cited are not involved in a thermal power plant operations, the quantum however varies time to time and place to place due to specific hurdles in operaion. Hence, it is necessary to be aware of the harmful effects of some of the constituents, if found, in excessive amount in the waste water to be discharged into receiving body. These are indicated as below, though the same may not be simply waste water borne from TPS: Permissible limits of these in water, where indicated, are generally taken as per drinking water standards.

Constituents	Effects
pH	Palatability, alimentary canal & eye irritation, damage of tissues, fish life & wild life propagation, irrigation & plant growth. Low pH corrodes the metal, keeps compounds of Al, Fe, Mn, etc. soluble if present in water as in case of ash pond decanted water. Similarly alkaline pH makes As, B, V, Sb, Cr etc soluble, and also at high pH precipitates hydroxides of Mn, Zr, Co, Cd, etc. if present in water.Desirable maximum limit as per WHO is 7.0 to 8.5.
Suspended matter	Gastro intestinal disorder, fish culture, bathing. Max desirable limit is 5 NTU turbidity.
Oil & grease	Gastro intestinal disorder, fish culture, bathing. Maximum desirable limit is 0.01 ppm.
Dissolved salts	Palatability, gastro intestinal irritation, fish culture, irrigation. Maximum desirable limit is 500 ppm, twice of chloride & should be greater than total hardness by 35 ppm.
Organic matter	Various types of diseases particularly when faeces, urine, bacteria, etc. available. Max. desirable limit as $KMnO_4$ (mg of O_2 absorbed) is 1 ppm.
Sulphate	Gastro intestinal irritation when combined with magnesium / sodium, cathartic action. Its max desirable limit is 200 ppm.
Chloride	Imparts salty taste, corrosion in hot water system, interference with agricultural & industrial processes. Desirable max limit is 200 ppm.
Nitrate/Nitrite	Carcinogen in animals stimulates algal & aquatic growth. Potential threat to infants for disease named 'methaemoglobinaemia' cyanosis, dangerous disease blue babies). The desirable max. limit is 50 ppm, although up to 100 ppm can be tolerated.

Constituents	Effects
Calcium/ Magnesium/ Sodium	Taste, corrosion / scaling in hot water system, cardiovascular problems in animals, irrigation by sodium absorption ratio. As per WHO, 150 ppm of magnesium is permissible with still lower if sulphate is present. Corrosion indication with absence of temporary hardness.
Iron	Taste discolouration, deposits, growth of iron bacteria. The desirable permissible limit as per WHO is 0.1 ppm (with max. limit 1.0 ppm), but even at 0.5 ppm definite complaint of foul smell, bitter taste, staining on cloth etc are noted. Water containing iron is often corrosive.
Manganese	Taste, discoulouration, deposits & turbidity, affects central nervous system & kidney, manganese poisoning known as "Manganessm" resemble 'Parkinson's disease'. Causes cramps & tremors. WHO gives 0.05 ppm Mn as highest desirable limit with max permissible 0.5 ppm.
Copper	Astringent taste, discolouration, corrosion of pipes, fittings & utensils. 0.1 ppm of copper is sufficient to increase considerably the rate of corrosion of galvenised steel and as little as 0.02 ppm will cause pitting of aluminium. Water having 1 ppm or less of copper can cause green stains on sanitary fittings.
Zinc	Astringent taste, opalescence and sand like deposits, causes high fever, malaise, dizziness, vomitting, diarrhoea initiation to gastro intestinal track, zinc toxicity in cereal crops leads to leaf chlorosis. Excess of 400 mg/g of Zn in soil may be toxic.
Fluoride	In excess of 0.9 - 1.1 ppm, can cause mottling of enamel of teeth. But even the above limits can cause dental decay when child takes such water during period of tooth formation.

Constituents	Effects
Inorganic ammonia	Growth of microorganism increases chlorine demand difficulties in chlorination. With phosphorus, it can lead to development of undesirable aquatic growths, toxic to fish, can be converted to nitrate and in the process can deplete oxygen resources. Danger of corrosion in pipes.Critical concentration could be anything depending on pH & temperature. Normal desirable limit is max.0.5 ppm with total nitrogen compound as 1 ppm.
Free CO_2	Damage in pipes due to corrosion, danger of bringing toxic metals into solution, taste and odour
Dissolved O_2	Growth of microorganism if the concentration is too low, (less than 5 ppm) formation of protective layer to metal is also hampered on low oxygen content, causing all the free carbonic acid of a non-aggressive water to be corrosive to iron piping.
Nickel	Human carcinogen in relation to pulmonary exposure.
Detergents / phosphates / surfactants	Taste and foaming, stimulates algal & aquatic growth, interferes with coagulation and in lime-soda softening. Max. desirable limit is 0.2 to 0.4 ppm.
Total hardness	Excessive scale formation, danger of dissolving heavy metals if the level is below 100 ppm (max. desirable limit as per WHO). The softer the water (very less Ca & Mg), the higher the incidence of cardiovascular disease in human being and corrosive to iron piping particularly when chloride is in imbalance form (Cl:hardness = approx.1:3)
Arsenic	Skin & respiratory cancer, auditory changes leading to total deafness. As per WHO arsenic should not exceed 0.05 ppm but levels up to 0.2 ppm are not known to cause any difficulty.
Cadmium	Toxic at high concentration to animals & aquatic life. Causes lung cancer, food poisoning through fish. Excess Cd may lead to disease characterised by rheumatic symptoms with intense pain in bones, replaces calcium in bones making them brittle. Recommended max. limit is 0.01 ppm.

Constituents	Effects
Mercury	Dangerous cumulative toxicant causing insidious, progressive chronic poisoning in mammals & fish. The max limit is generally 0.00005 ppm
Lead	Detrimental effect to human health with actute toxicity in children including mental retardation. As per WHO, max. limit is 0.1 ppm and 0.3 ppm when in contact with lead pipes.
Chromium	Staining on clothes & skin. Max recommended limit is 0.01 ppm
Chlorine	Harms fish & wild life propagation. Though 0.2 ppm residual chlorine is good enough without anyone noticing its foul smell, but sometimes even 0.05 ppm can give foul smell, if this chlorine reacts with some grease on a tap or drinking utensils. An amount of 0.001 ppm phenol may also react with chlorine to form an objectionable taste of 'chlorphenol. Manganese is easily oxidised by chlorine.
Temperature	Harms fish & wild life propagation. Should not be more than +/- 5 ^{0}C within 10 metre radius of high temperature water mixing point, otherwise at ambient should not be greater than 35 ^{0}C.
Dissolved oxygen	Fish life & foul smell. Minimum desirable limit is 5 ppm.
Hexachloride/ petrochemicals/ phenolic compounds cyanide / selenium	May be related to the development of cancer, also may cause taste & odour problems. Max desirable limit is 0.02, 0.01 & 0.001 respectively. Max. desirable limit for cyanide 0.05 ppm and for selenium is 0.01 ppm

There are various other rare elements like berrylium, thallium, titanium, etc. toxic to human health and nature. Hence, regulations are needed to check pollution of regular receiving body water through the waste water emanating from any industries, including from thermal power plant.

Major sources of waste water generation in TPS

1. Cooling water from condensers & auxiliaries (using common ducts) and
2. Overflow from ash pond. Other sources of waste water are:
   - Boiler blowdown
   - Water treatment plant discharge including from DM plant
   - Run off due to rain/dust suppression from water from coal handling area
   - Waste water from oil handling area
   - Floor drains
   - Waste water from few auxiliaries cooling having separate closed. Cooling water circuit also, intermittently. Waste water is associated with commissioning, start up, shutdown, etc. including chemical cleaning of the plant.

Pollution Impact & Measures to Contain Cooling water discharge:

If the power plant adopts an open cycle system, the entire water quantity of the order of 30,000 m^3/h for every 200 MW will be led into the water body at a temperatue up to 5-7^{0}C higher than the ambient. In a closed cycle of circulating water system however, a certain percentage of water has to be regularly blown down in order to maintain the specific concentration of dissolved salts in the water of the closed cycle.

Generally, chlorine and the temperature of the discharged water from a TPS are the potential danger for the receiving body. Chlorine reacts with organic matters of the water (which are found more often in the source water contaminated with sewage or industrial effluents) to form chlorinated phenols & hydrocarbons, which are priority pollutants. Chlorine or the reaction product with ammonia (if naturally found in the source water) as chloramines are also common chemical threat to fish & lower hydra. Hence, removal of organics & other contaminants including sewage & industrial effluents is necessary before chlorination, thus demanding for use of clarified water in the circulating cooling water. Further to avoid threat of fish & other aquatic life from chlorine & chloramines, it is necessary to monitor these contents in the waste water so as not to exceed their level more than 0.2 ppm residual as TRO (total residual oxidant). Earlier, the recommended level for these had been found as 0.5 ppm, but now reduced in view of adverse effects noted.

Under TPS practice, chlorination to the incoming water on once through cooling system is not continuous, and the residual values in the outgoing water to the receiving body do not generally exceed more than 0.1 ppm as TRO. Similarly, under cooling tower cooling system, practices are to use clarified water again with intermittent dosing of chlorine, and the blowdown from cooling tower is not being done when chlorination is going on. Thus, cooling water waste from a TPS is not generally harmful with respect to chlorine, when the same is discharged to the receiving body. Few TPS even use the cooling tower blowdown water to use for making ash slurry to be discharged to ash pond.

Thermal pollution is another problem associated with the cooling water particularly when the cooling system is once through & not through cooling tower. Excessive temperature of water to be discharged into the receiving body like lake, river, sea etc, affects physical, chemical & biological characteristics of water bodies. Though, some studies in the country have indicated that there is no impact on fish life up to about 10 ^{0}C temperature rise, unless there is sudden discharge of thermal effluents, and hence no significant change is expected in seasonal trend of various groups of naturally found algae, photoplankton, etc. yet limits of acceptable levels of thermal discharge to receiving body should be defined by either of the following:

- Temperature rise at a point
- Average temperature rise over a surface area/stream cross section
- Rate of temperature rise

Hence, an average rise of 0 ^{0}C relative to normal ambient is generally prescribed to maintain trout stream conditions, whereas local rises of as much as 5 ^{0}C should be acceptable for streams, for which differential temperature across condenser should be limited to 10^{0}C. Thus, pollution control authorities do not allow a water to come into the receiving body at more than 5 ^{0}C of its own temperature. Thermal pollution can change the eco system of the receiving body in the following way:

Dissolved oxygen: Saturated concentration of dissolved oxygen with ambient air in general is necessary for maintenance of aquatic life, but the same decreases with increasing temperature of water. At the usual reference temperature of 20^{0}C, equilibrium oxygen concentration is 9.2 mg/l. Increased temperature has further an undesirable effect on reaeration rate.

Algal response: Production measured as cell mass or oxygen produced will increase with increasing temperature to a maximum. Though magnitude of maximum productivity & associated temperature are a function of nutrient supply, illumimination & prior conditioning, above the optimum temperature, productivity exhibits a rapid decline. Along with, temperature rise there is tendency for one group to be displaced by another, viz., chlorophyta (green) to cynophyta (blue green) to others (brown, red or yellow green) and hence temperature response by an algal community is essentially a continuum.

Bacterial assimilation: This is a principle means by which organic compounds and ammonia are removed from natural water, which involves both respiration (also called maintenance) and reproductive growth of different bacterias.

Hundreds of enzyme catalysed reactions together with simple and activated transport are implicated in these cellular activities. However, there is little change in this activity provided the temperature variation is between 10 and 37.5 ^{0}C.

Higher organisms (cf: Fish): Like microbial population, macrobiota also respond to temperature variations completely. Fishes have upper & lower thermal tolerance limits, which vary considerably from species to species. In addition, reproduction may not occur at temperatures which are otherwise consistent with survival. Failure to reproduce may be due to lack of conditions suitable to spawning or limited survival of young fish. Finally, there is a complex interaction among temperature, oxygen concentration & toxic chemicals. As upper and lower tolerance limits are approached, fish are significantly less resistant to reduced oxygen concentrations & dissolved pollutants.

The following are the general guidelines as issued by the pollution control body in respect of waste cooling water from a TPS:

Condenser Cooling Water Waste (once through cooling system):	Temperature	not more than 5 ^{0}C higher than intake water temperature
	pH	6.5 – 8.5
	Free available chlorine	0.5 ppm
Cooling Tower Blowdown:	Apart from the above three items, viz., temp, pH & chlorine, the following in addition are required:	
	Zinc	1.0 ppm Total ban
	Chromium in any form	5.0 ppm
	Phosphate (total) Other corrosion inhibiting materia	Limit to be established on the basis ofpollution control authorities

Thus, under a TPS practice, temperature is not generally a problem when water is being discharged from the cooling tower, but where there is once through cooling, little higher temperature up to 7 ^{0}C rise have been noted. Under the circumstances, efforts are on to make the discharge system to enable rapid mixing in a small mixing zone, so that water is cooled down before discharge, otherwise provision of cooling tower to be made mandatory.

In case of sea water cooling, temperature rise is not a concern as discharged warm water mixes with large quantity of sea water. Still, it would be better particularly with back sea water to stagger the discharge into atleast four or more segments at different points of sea. Warm water will thus mix the sea water without disturbing the marine life.

Where antiscalants or anticorrosive chemicals are added in the circulating cooling water system with cooling tower, the same have generally not been noted to cause concern, up to a concentration increased by COC (cycle of concentration) of 5.0. Anti-corrosive chemicals involving chromium are nowadays not even manufactured. Additional precautions are to use the cooling tower blowdown water in making ash slurry for disposal to ash pond, and also to make use of this water in flue gas desulphurisation unit and finally attemting to treat the water suitably along with other waste water for own use of TPS, as discussed under reverse osmosis treatment programme in 'water treatment chemistry'chapter.

Boiler Blowdown Water: Boiler blowdown is usually of high quality and very small in quantity (25 -30 m^3/h for a 500 MW TPS) and does not contain any components beyond allowed limits. Only concern is that this water has high alkalinity and has very high temperature. The following are the maximum limiting concentrations, needed to be maintained as guided by Pollution Control Bodies:

Contents	Max. Limits
Suspended solids	100 ppm
Oil & grease	2.0 ppm
Copper (total)	1.0 ppm
Iron (total)	1.0 ppm

In almost all situations, boiler blowdown under TPS practice is principally a source of little soluble sodium salts and ammonium hydroxides and in that way the above contents are found always well within limits. Only concern is for high temperature & pH which is dealt by leading this water to a flash tank, from where the steam portion is being utilised again to make a part of feed water and the remaining water portion is conditioned with respect to temperature & pH by diluting, and as per quality, the same is either used or discharged to waste through DM plant neutralisation pit after its cooling and neutralisation.

Boiler & other metal cleaning wastes from the boiler, though are occasional, but contain objectionable contents which need to be suitably treated before discharge. Generally, for cleaning alkali like Na_2CO_3/NaOH and acid like HCl /HF are used with certain additives like detergents/surface active agents. They are first neutralized with suitable chemicals like acid and alkali and then suitably diluted before discharging mostly through ash slurry discharge into the ash pond. In absence of ash slurry system, these wastes are generally tried to be contained in the power plant premises itself in the concentrated solid form for solid disposal or mixed with other wastes suitably to be further treated mostly for own use of TPS for the purpose even desalination plant is employed. For the conditioning purposes of boiler, predominant use of hydrazine are there, which when obtained as waste is neutralised with bleaching powder to obtain finally as solid wastes for discharge. Similarly, if HF acid is used as cleaning agent, the same is neutralised with lime to obtain as solid wastes. These activities of cleaning in a TPS is generally not more than once in a year, and since amount of cleaning agents are also not huge, no separate plant is therefore needed to treat them separately.

Water treatment plant / DM plant effluent:

The pretreatment section of water treatment plant has sludge slurry for discharge, but these are generally either discharged along with ash slurry to ash pond or concentrated in small pits & dried in the plant area itself for solid disposal. The decanted water if any available in this way is led to DM plant waste water for dilution.

The DM plant waste water generally contains acids and alkali and backwash water from filters. Except pH of 6.5 to 8.5 there is so far no specific other guidance from pollution control body in this connection. The acids & alkalis obtained in the waste are properly neutralised in a suitable pit to obtain the recommended pH before discharge to a receiving body. However, the total dissolved salts content in this water, which also contain heavy metals like iron, aluminium, manganese, copper, zinc, etc. apart from usual calcium & magnesium and acid & alkali as regeneration waste, may also figure sooner under the guidance of pollution control body, and then this waste water will additionally require adequate dilution or suitable treatment before its discharge.

Waste water from oil & coal storage handling area:

As per TPS practice, oil spilled during transfer in and out of storage tanks are washed out by water and accumulated in a dip pit, from which oil is separated out by oil separator and suspended solids are allowed to settle in the pit along with waste water to be further pumped out. Similarly in the coal storage area, waste water generated during dust suppression or run off due to rain is led to a pit first or settlement of dust particles and the decanted water is then pumped out for disposal. This waste water is however very less in a TPS and the quality does not pose a threat against pollution norms. Yet, where separate waste water treatment is envisaged, this water joins the plant routed through neutralisation pit of the DM plant.

Other waste water:

a) Pits of smaller sizes are provided in different areas for collection of waste water including of floor washing & then pumped to neutralisation pit of DM plant, where it is further oxidised by air & mixed with other effluents. This minimises both suspended & dissolved solids before discharge.

b) Ash pond discharge water:

The waste water available from ash pond decantation is generally 70% of ash slurry water input to the ash pond. The coal ash in most of the power stations are discharged out of the plant area to a large ash pond in the form of slurry (ash:water = 1:5), the water being drawn either directly from source or through cooling tower water blowdown or both. A 210 MW unit using coal of calorific value 3500 kcal/kg and ash content in the range of 40% produces ash at the rate of approx. 60 T/h. Normally, the ash is dumped in the form of slurry in the low lying areas where about 10 to 15 metre of depth is available, which helps in reclaiming the land. The ash settles down and fills up the area while the decanted water from the pond flows out into neighbouring fields, drains, river, canal, etc. If sufficient ash particles are found in this decanted water, there is likelihood of ecology disturbance due to the flora and fauna imbalance in the water. Ash particles are different in character to the natural soils and the nutritious food regain of various organisms living in water are destroyed.

If such water is led to agricultural field, they leave behind a white crust on the field making it non-fertile. Department of environment has therefore laid down the following limits for ash pond decanted water:

pH	6.5 to 8.5
Suspended solids	100 ppm
Oil & grease	20 ppm

Survey of the performance of coal fired thermal power stations, however, indicate that a well built ash pond with variable spillway and rockfill embankments with sufficient retention time leave behind very little suspended solids as low as 10 ppm and pH 7.2 to 7.5 with oil & grease virtually nil and thus there is no difficulty at all in maintaining the above quality standard of the decanted water. Possibility of carryover of more ash particles in the decanted water is expected only after filling up of the pond with ash (usually after 5 to 7 years, after which new pond is taken into service) giving rise to channeling flow in the pond instead of holding lagoons. Carryover is also noticed due to excessive natural disturbances like wind. But investigations reveal coal ash to contain certain toxic metals with one typical data as below:

Arsenic	290 ppm	Berrylium	50 ppm
Cadmium	9 ppm	Nickel	520 ppm
Lead	550 ppm	Thallium	35 ppm
Selenium	225 ppm	Zinc	335 ppm

Environmentalists therefore have more fear that ash ponds in the long run could impurify the surrounding ground water by the way of ash leacheates and hence they recommend to have suitable lining of the ash pond, which is highly cost- effective, sometimes technically not feasible also. Hence, location of ash pond is required to be so chosen that soil has little permeability of water to contaminate the ground water. Soil should have also less absorbance and more ion exchange properties. Again, to make the pond decanted water almost free from ash particles, it is necessary that the fly ash at TPS is blended with bottom ash and clariflocculator sludge for easy settling in the pond and the decanted water is reused by TPS itself for ash sluicing by recycling method. Otherwise, suitable clarification/filtration plant is required to purify the ash pond decanted water from ash particles. Maximum utilisation of fly ash should also be explored with its dry ash collection & disposal system, which is also needed in view of large amount of land (about 500 hectares for a 1000 MW TPS) requirement. With the rapid depletion of natural sites i.e. low lying waste land, acquisition of fertile land would become necessary for future TPS, causing in due course severe ecological imbalance and related environmental problems. Viable fly ash products could be as below:

i. Calcium silicate bricks
ii. Clay fly ash bricks
iii. Sintered fly ash aggregates
iv. Cellular concrete
v. Portland pozzolana cement
vi. Admixture with cement

Based on this, consumption of fly ash has already started by setting up suitable industries, still the production of fly ash has been too much in comparison to demand by these industries. Hence, all new TPS of the country now generally have both wet and dry disposal system of fly ash, few of them disposing bottom ash only in the form of slurry and mixing fly ash only during emergency.

Seepage of ash pond water to contaminate ground water is, however, not yet widely proven and requires detailed critical analysis site to site in respect of type of soil & bedrocks, the quality of ash and the movement of ground water in Talchir Aquifer. If the nature of ash, as usually, found is of cementing, the cementy clayey layer at the bottom of ash pond will prevent seepage to the ground, although the ash and the earth below may be able to retain about 50% of seeped water. The coefficient of permeability of clay has generally been found 0.000001 cm/sec proving practically no percolation of water induced by gravity in clays. The seepage movement is further complicated by the presence of ground air, which is displaced in part by the downward seeping water and completely by a rising water table. The penetrating power also depends on pressure, heat & corrosive chemical composition of earth. There could be other variables also governing the seepage. Thus, simply fear should not be expressed that ash pond is a source of inducement of ground water quality deterioration, but it is necessary to come out with more facts & figures. However, if adequate vegetation of suitable herbal family are grown in the ash pond, the seeped water retained by the clayey layer at the bottom of the pond would be continuously utilised by the plant and even the minute danger of percolation to the ground water could be avoided.

Based on the above, it is thus seen that potential threat to environment from the waste water as available from a TPS (coal based) is mainly from water treatment/ DM plant, and cooling tower blowdown (if not used for ash sluicing), which must be treated suitably before its discharge to the environment or to the receiving body or for own use for the TPS. One such programme of treatment with solid waste disposal is illustrated in figure 1 below. Other waste water, viz., ash pond decanted water available on regular basis abundantly, and other available occasionally in small amount do not pose so many problems in view of care taken of during design stage of the coal based power plant.

Yet, in view of scarcity of water almost at every site and the pollution control norms going stringent day by day, it is imperative that the management explore the possibility of both the discharge and recycle of waste water in the best way.

Waste Water Treatment:

Treatment of waste water for discharge and for recycle is a complementary approach to waste water management. Under TPS practice, either one or both are followed, which vary site to site. However, treatment programmes come into two parts:

a. Ash pond decanted water.

b. All other waste water in the plant area.

In view of general performance of ash pond decanted water invariably meeting the requirement of pollution control norms and some power stations even recycle this water for sluicing the fresh ash generated. Most power stations do not have any treatment programme for this waste water. Where the treatment is found essential (generally with old ash pond) separate treatment programme for this is envisaged which include suitable clarification and filtration plants in series and the plant located generally at ash pond site itself, away from power station site.

All other waste water is collected in the neutralisation pit of DM plant, equalisation tank or in stilling chamber as shown in figure below (typical of 100 MW TPS) and is treated suitably as indicated to discharge clean water to the receiving body and dry sludge / solid disposal in the form of cake separately.

Since suspended solids/sludge is the main concern for treatment, the figure also illustrates the typical solids loading of different areas with the final one. The waste water from ash handling plant and the sludge from clariflocculator have not been shown here. The same are supposed to be discharged separately to ash pond. For making ash slurry, in some TPS, some portion of cooling tower blowdown water is also utilised. In view of water being sufficiently diluted in equalisation tank, other ingredients have been found well within the norms of waste water quality suitable for discharge, as guided by pollution control boards.

In some power stations, due to water scarcity, waste water is treated using reverse osmosis plant for reuse.

Pollution control against ash

Ash generated in power plant is about 30 - 40% of total coal consumption and hence the system is required to handle ash for its proper utilisation or disposal. Fly ash (around 80% is the value of fly ash generated). Bottom ash (bottom ash is 20% of the ash generated in coal based power stations).

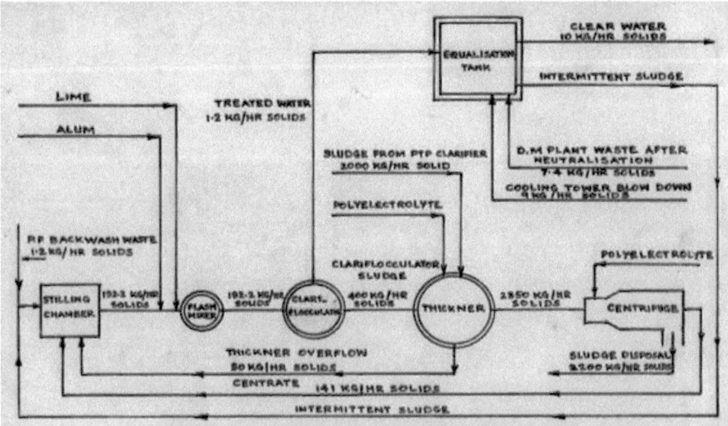

FLOW DIAGRAM EFFLUENT TREATMENT PLANT FIGURE 1

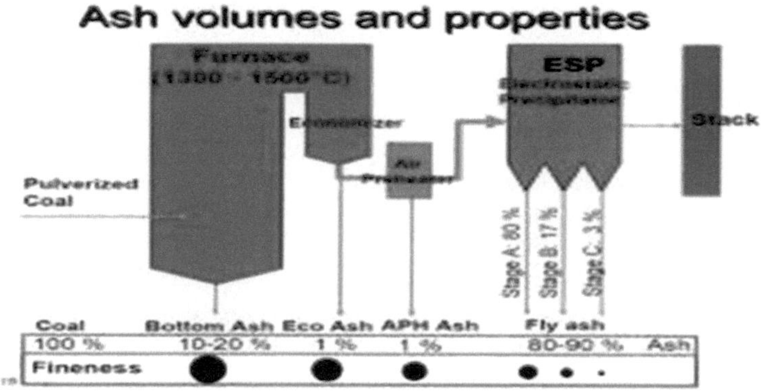

Fly Ash - Ash generated in the ESP which gets carried out with the flue gas is generally called fly ash. It also consists of air pre-heater ash & economiser ash (2% of the total ash content). Ash generated below furnace of the steam generator is called the bottom ash.

Treatment of Effluents

Recovery and Reuse of Water

System Description

The ash handling system handles the ash by bottom ash handling system, coarse ash handling system, fly ash handling system, ash disposal system up to the ash disposal area and water recovery system from ash pond and bottom ash overflow. Description is as follows:

A. **Bottom Ash Handling System**

Bottom ash resulting from the combustion of coal in the boiler shall fall into the overground, refractory lined, water impounded, maintained level, double V-Section type/W type steel-fabricated bottom ash hopper having a hold up volume to store bottom ash and economiser ash of maximum allowable condition with the rate specified. The slurry formed shall be transported to slurry sump through pipes.

B. **Coarse Ash (Economiser Ash) Handling System**

Ash generated in economiser hoppers shall be evacuated continuously through flushing boxes. Continuous generated economiser slurry shall be fed by gravity into respective bottom ash hopper pipes with necessary slope.

C. **Air Pre-heater Ash Handling System**

Ash generated from APH hoppers shall be evacuated once in a shift by vacuum conveying system connected with the ESP hopper vacuum conveying system.

D. **Fly Ash Handling System**

Fly ash is considered to be collected in ESP hoppers. Fly ash from ESP hoppers is extracted by vacuum pumps up to intermediate surge hopper cum bag filter for further dry conveying to fly ash silo.

Under each surge hopper, ash vessels shall be connected with oil free screw compressor for conveying the fly ash from intermediate surge hopper to silo. Total fly ash generated from each unit will be conveyed through streams operating simultaneously and in parallel.

F. **Ash Slurry Disposal System**

Bottom ash slurry, fly ash slurry and the coarse ash slurry shall be pumped from the common ash slurry sump up to the dyke area which is located at a distance from slurry pump house.

Chapter - E

CASE STUDIES ON PROBLEMS FACED BY FEW TPS

(This chapter deals with real problems faced by different TPS of the country as studied by the author, with remedial measures. It could further be helpful in understanding the importance of water chemistry for good performance of TPS)

Case Studies – Water Chemistry and Its Impact On TPS

The author had occasions to visit different TPS of the country as roving monitoring team member of CEA and noted certain trouble shooting part in water chemistry regime, resulting to loss in power generation. When the problems were duly corrected, the performance of the power station improved. The survey is therefore self indicative how water chemistry can have a good impact on power plant generation.

Case study 1

There have been severe failure of pipe lines & equipments of auxiliaries cooling water system including discharge pumps in a TPS of 2 x 60 Mw, in spite total soft water under circulation in the system for cooling. The generating units have to be finally taken out of service for few days. When the arrangement of soft water blending with filtered water in a fashion of making chloride vs carbonate hardness in the ratio of 1:3 in it, was done under circulation in the newly replaced pipelines & pump, the system worked well without any trouble.

Case study 2

Under soft water cooling system of condenser of 4 x 60 MW station there has been considerable choking of condenser tubes and the generating unit was not able to achieve optimum generation. The deposits as found in the condenser tubes were mainly corrosion products contaminated with siliceous particles. The leakage of hardness from the softening plant was 2 ppm. This when increased to 5 ppm, the circulting water did not create further deposits of corrosion products in the condenser tubes. Under the same power station of 2 x 12 MW units, the condenser tube deposits contained mostly aluminium oxides deposits, the cooling water being clarified water under cooling tower facility. When attention was paid to have good clarified water controlled alum dosing in the clarifier, the problem of deposits was solved.

Case study 3

There has been very high silica content in boiler water of a 2 x 120 MW units (0.5 - 1.0 ppm) continuously in spite of make-up water to boiler having less than 0.01 ppm. When make-up water was analysed for colloidal silica, it was found to be 0.02 ppm. An ultra filtration plant was then installed after mixed bed exchanger in the DM plant, which was able to give make-up water to boiler always less than 0.02 ppm as total silica, and when make-up water quantity was also reduced to 3% in place of existing about 5%, the silica content in boiler water continuously remained less than 0.2 ppm with almost no blowdown to boiler. Yet in another power plant, where raw water contained more colloidal silica in the range of 2 - 3 ppm, it was preferred to get it removed at the pretreatment stage itself by suitable alum and polyelectrlyte dose in clariflocculator followed with both slow gravity filter and slow pressure filter of dual media & having facility of further polyelectrolyte dosing arrangement, the filtered water itself was practically free from colloidal silica.

Case study 4

Efficiency of the generating units of different sizes in a TPS had been too poor due to high demands of make-up water (almost 16%) resulting to severe disturbance in heat balance. The study team meticulously pointed out the leakages of the make-up water and steam at various points in the circuit and finally were able to stop the undue leakages and restore the efficiency. The proposal for installation of additional streams in DM plant to meet the demand of make-up water was also then scrapped.

Case study 5

During commissioning of a new DM plant of 75 m^3/h. The mixed bed resins were clumped together in the exchanger vessel and the DM plant was not able to produce DM water for boiler commissioning. Under the pressure, the DM plant supplier was almost ready to reject the resins with fresh resins fill in the mixed bed exchanger vessel. At appropriate time, it was suggested to keep the vessel filled with 2% NaOH solution for two hours and then drained & backwashed properly. Both the cation and anion resins were separated for their effective regenerations, and thus the DM plant stream was able to produce DM water without rejecting the resins and save huge money loss apprehended by the representative of the plant supplier.

Case study 6

Not only early damage of cation resins due to deposits of calcium sulphate on it, but also damage in lead lining in the pipelines in the sulphuric acid dosing system of cation resins regeneration system caused shortage of make-up water to boilers. The root cause was found to be changed raw water quality towards more inferior than the regeneration system designed for. In power station, the total regeneration system had to be changed to HCl based regeneration system, and in another low capacity DM plant, due to cooling arrangement at mixing point of sulphuric acid with water, and proper two stages regeneration system through well designed measuring tanks sustained the problems.

Case study 7

Counter current regeneration with air hold down from above the resin bed has proven damage of cation resin functional groups, due to air oxidation, particularly when the regenerated resins were kept idle for more than a week, and kept as spare. This when changed to water hold down pressure during regeneration, resins were found alright with their exchange capacity even after a month of regenerated resins kept idle for spare. In yet another station, frequent breakage of middle collectors due to canty lever action, the exchanger unit was used to be regenerated by reversing the acid inlet path in the same exchanger vessel i.e. on co current basis with backwash during every regeneration without getting the due performance. This led to a feeling of having a suitably modified middle collector or thinking of counter current mode of of regeneration only in case of strict need of low leakage less than 0.5 ppm sodium or less than 0.1 ppm silica at respective exchangers, otherwise think of only co current regeneration. For anion unit, a second thought of achieving less than 0.1 ppm silica by warm caustic soda (50^0C) regeneration seemed preferable in place of going for counter current regeneration, as in the later case, chances of polymerisation of silica on resin bed during each stage of regeneration are more?

Case study 8

Condenser cooling system did not exhibit $CaCO_3$ deposits in few power stations in case of cooling water having clarified water under open recirculating system with cooling tower. Perhaps CO equilibrium with $CaCO_3$ & $Ca(HCO_3)_2$ in a cooling tower is different than without cooling tower. True equilibrium is never achieved in a cooling tower due to presence of rampant CO above the water surface in cooling water basin not able to escape from the tower to the atmosphere. In these power stations, predominant deposits in condenser tubes were found to be corrosion products only contaminated with dust/silt. In a cooling tower in one of the power stations having sea water cooling exhibited too many live barnacles deposits all along the cooling equipments in touch with sea water incuding condenser tubes and cooling tower basin. These were very difficultly controlled after a series of experiments with different types of biocides including alkyl dimethyl benzyl ammon. Chloride, sodium hypochloride and low molecular weight anionic polymer, etc. Gaseous chlorine was not possible to be used here due to environmental problems. These barnacles were supposed to be imported ones, as pumping station of sea water for cooling system was from a point near to, where generally many ships were anchored and cleaned for ship bottom deposits of clams, mussels and barnacles. From the same sea, in another power station but withdrawal source far away from this ships' anchoring point, and having once through cooling system without any cooling tower did not exhibit any such deposit problems. In terms of biocides, only chlorine generated through sea water itself through electro chlorination, was in use. These barnacles & mussels also have led to calcium carbonate scale in the titanium condenser tubes, which were later cleaned by sulfamic acid. When barnacles/mussels and brown algae were controlled, no deposits of calcium carbonate were found in the condenser tubes.

Case study 9

Inadequate pre-commissioning cleaning of boilers have led to steam impurity conditions for a very long period causing even breakage of turbine blades due to turbine deposits and loss of power for months together for a 52 MW power station. Sufficient amount of mucs remained in the boiler system to cause disappearance of useful phosphate in the boiler drum for due solid alkalisation of boiler water. Frequent start ups and shutdowns of boiler with very low phosphate dosings and with heavy blowdowns of boiler together with mechanical cleaning of boiler drum and condenser hot well at each shutdown time including finally turbine wash could finally restore gradually the steam purity and the power loss. Too many steps for boiler water treatment were suggested by different experts, but finally low phosphate dose of 2 ppm as PO_4 under congruent treatment programme of phosphate with minimum blowdown continuously was helpful in achieving the steam purity. But yet there have been again turbine blade failures after two years only. Chemical treatment programme remaining the same, turbine blade metallurgy was changed this time, and the turbine is now running well with much cleaned surface since two decades approx. as found on physical inspection.

Case study 10

In another power station of 140 MW, turbine failure with blade breakage and turbine disc cracking were noted after its approx. one year of its commissioning. Although failure causes were primarily viewed as metallurgical failure, but disc cracking was attributed to organics as found in steam, since boiler make-up water contained more than 200 ppm of total organic matters, and the boiler averagely ran with more than 10% make water requirement.

Case study 11

Purity of condensate & feed water was not clear, since the cation conductivity as determined by laboratory cation column (online metre not available) were always found more than 1.0 micromhos/cm, in spite of base load operation of unit continuously and make-up water having 0.2 micromhos/cm conductivity. In one power station of 140 MW units, morpholine along with ammonia & hydrazine were dosed in the feed water, while in the other of 110 MW units, cyclohexylamine was added along with others. When use of morpholine/cyclohexylamine are given up in all units, the cation conductivity as measured by laboratory, stood to 0.3 to 0.5 micromhos/cm only, giving impression that feed water-steam cycle purity were considerably acceptable. This gave belief that morpholine/cyclohexylamine probably formed some complex variety with corrosion products which could not be exchanged by the cation resin of the column. For other reasons also with respect to their volatility & breakage at high pressure & temperatures, along with more copper pick up, use of these in high pressure & temperature boilers were later given up by almost all power stations.

Case study 12

Both boiler drum & turbine blades of a 60 MW per unit power station contained severe deposits of copper along with iron oxides which could not be cleaned effectively and the turbine had to be deaerated. But after this when ammonia dosing were optimised & performance of deaerator & condenser improved with respect to dissolved oxygen content of feed-water-steam cycle, the copper deposits in further operations were found no where.

Case study 13

Unawareness of due water chemistry programme have led many power stations in trouble including their generation loss, few of which can be cited as below:

a. In one of the power station, condenser leak was declared based on the hardness found in the condensate and accordingly the unit was shutdown for detecting leak & its repair. When the leak could not be detected even after its search during its three day shutdown, the source of hardness was revealed due to one idle HP heater taken into service (idle since three months), which led to increase of corrosion products and finally hardness detection (due to iron) by usual Eriochrom Black T method. Actually, cation conductivity determination of condensate at that time (compared with the previous one) could have been the best option, before declaration of condenser leak. Neither online nor laboratory was equipped with this.

b. In another power station the condenser leak was declared based on high silica content in the condensate, though it remained under watch without its shutdown, in view of silica content sometimes coming down to normal value also. This variation led to a belief that the leak is minor and the leak point intermitanly closing & opening with some solid particles. The same was finally detected due to steam impurity, both due to carryover from boiler drum and the turbine wash under low load conditions. Intermittent blowdown to boiler with increase of MW varied silica and sodium content, creating confusion of leak point opening & closing.

c. In a number of power stations, boiler water pH was not able to be maintained in the required range of 9.3 - 9.5, in spite of high dose of tri sodium phosphate due to presence of corrosion products in the boiler water and subsequent hideout of phosphate. The blowdown to boiler was not increased on the false presumption that blowdown had been to control silica content only. On increase of blowdown including sometimes the intermittent too and low phosphate dosing, the system gradually restored to desired function.

d. Again in a number of power stations, boiler feed water pH and dissolved oxygen contents could not be controlled as per desired value of 8.8 - 9.2, in view of same tank being used to dose hydrazine & ammonia as a mixture. When individual tanks and pumps were used for these, and the dosing amounts were made proportional to the quantum of make-up water, the system worked well. Also one extra dosing point (apart from economiser inlet) at condensate pump discharge was made to take care of very high make-up water requirement at the cold start of the generating unit, and to be operative during this time only.

Case study 14

Unskilled workmanship unknowingly has created incidental problems causing delayed generation time as below:

a. In view of pre-commissioning acid cleaning of boiler with HCl taking too much time in bringing it up and still feeling dangerous from corrosion point of view particularly at tube bends containing deposits or at low flow conditions in the system, and also the superheater portions remaining to be separately cleaned, one of the power station opted for HF acid cleaning (though HF is quite dangerous to handle), followed with hydrogen peroxide passivation. But as ill luck would have it, after cleaning boiler & economiser contained lot of calcium fluoride deposits, delaying schedule of generation too much more than the originally it could have been with HCl cleaning. It was learnt that the economiser got sufficient sea water entry at unloading point at sea shore, which was not revealed to site personnels.

   Thus, the experience of HF acid cleaning in India could not be known full proof. However, to avoid the hazards of acid cleaning (HCl/HF) and to expedite the pre-commisioning schedule, these days high pressure boiler tubes and drums are generally made of cold rolled steels, for which alkali boil out sometimes followed with citric acid are found sufficient after due passivation. Post commissioning cleaning procedure are however adopted on the basis of deposits, if any found, otherwise alkali boil out with due passivation with hydrazine/ammonia is followed.

b. In another power station of 110 MW units, when the unit started loading as usual, the boiler water silica was found too high to give signal for steam entry to turbine. Even after two days of operation of boiler with sufficient blowdown, when boiler water silica was still in unacceptable positon, the boiler was brought down, and the drum opened for inspection. Unamazingly, in the middle of the drum, a pyramid type of structure around one & half feet high was found fully siliceous in nature. This was found to be powder of fire bricks, few of which were taken inside by the workers, probably for sitting purpose and forgot to take them out before boxing the drum. In yet another case a single shoe was found in hot well, which delayed the condensate cation conductivity to normal value.

Chapter - F

GUIDELINES ON WATER TESTING METHODS

(Describes testing methods in brief with updated technologies on all forms of water relevant for a TPS operation, along with principle of testing methods and the reasons for undertaking the tests)

Guidelines on Water Testing Methods

Titrimetric Method

Alkalinity:

In natural water, the alkalinity comes from the presence of bicarbonates and carbonates and to express the alkalinity, phenolpthalein and methyl orange indicators provide standard reference points, which change colour at pH 8.3 and 4.3 respectively.

100 ml of water sample at room temperature is titrated with an N/50 H_2SO_4 to the Phenolphthalein end point (pH 8.3) and then to methyl orange end point (pH 4.3). The phenolphthalein end point measures the hydroxide ions and half of the carbonate ions and methyl orange end point, the other half of the carbonate ions and bicarbonate ions.

Calculation:

Multiplying the titre value by 10 and the factor of the acid (if any), together gives the respective alkalinities (called as P & M alkalinity). Further, from P & M alkalinity, bicarbonate, carbonate & hydroxides can be calculated as below:

Alkalinity	Bicarbonate	Carbonate	Hydroxide
P = Nil	M	Nil	Nil
P = < 1/2M	M - 2P	2P	Nil
P = 1/2 M	Nil	2P	Nil
P = > 1/2 M	Nil	2 (M - P)	2 P - M
P = M	Nil	Nil	P

Precautions:

1. If the water sample is more turbid, it is better to measure a filtered sample for test. However, the P-alkalinity must be determined on the unfiltered sample.
2. For direct measure of hydroxide alkalinity from the phenolphthalein end point, barium chloride needs to be added prior to the titration to precipitate the carbonate ion from solution.
3. Use sodium thiosulphate before determination of M-alkalinity, if water is chlorinated. The end point in this case using methyl orange will be smoky grey. It is better to dechlorinate the water before testing for alkalinity and use phenol red indicator.

NOTES:

a. In place of methyl orange indicator, mixed indicator (composed of bromo cresol green & methyl red) giving a very sharp end point can also be used. 0.2 g of powdered bromocresol green & 0.04 g of powdered methyl red in 100 ml of chemically pure 80% isopropanol. The solution should not be filtered, but allowed to stand a day or two until the slowly soluble indicator are completely dissolved. Use 2 - 3 drops of this indicator per 100 ml sample of water.

b. 1.25 g of phenolpthalein in 250 ml of 50% isopraponol and then neutralised with N/50 NaOH solution. Use three drops of this indicator per 100 ml of water sample.

Acidity:

Acidity in water samples can be measured in two ways.

a. Free mineral acidity, having a pH of less than 4.3 and

b. Total acidity, having a pH of less than 8.3.

Free mineral acidity (FMA) is not usually found in natural water, but under water treatment practices these are found in decationised water having presence of sulphuric, hydrochloric and nitric acids. FMA however does not include carbonic acids if found present. Amount of sulphate, chloride, nitrate, fluoride, etc. computed to equivalence to their respective acids together is called equivalent mineral acidity (EMA), which is also approximately equal to FMA.

In non-polluted water, the total acidity of the water is generally caused by dissolved carbon dioxide and the pH is between 4.3 and 8.3.

For testing purposes, 100 ml water sample is titrated against N/50 NaOH using mixed indicator, and another 100 ml sample using phenolphthalein indicator, and respective acidities are calculated by a multiplying factor of 10 to give the contents as ppm.

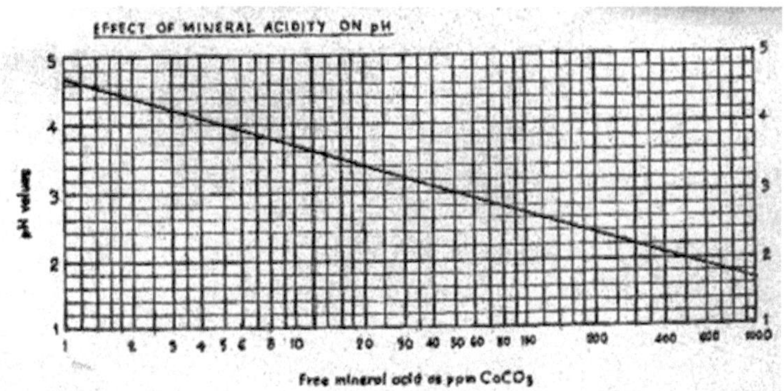

Total Hardness:

Total hardness in a sample of water is generally taken as the total amount of calcium and magnesium present in it, but the same can also be found intermittently associated with iron, aluminium, zinc, manganese, etc. if found considerably in the water and the testing procedure is adopted with same hardness indicator and buffer solution as outlined below:

The testing is done by titration method using EDTA solution as titre in presence of ammonia buffer of pH 10 and eriochrome black as hardness indicator as followed:

EDTA Solution: Dissolve 3.85 g of sodium salt of EDTA in 800 ml of pure water. Add 0.5 g of sodium propionate as preservative. Adjust pH 6 to 7 by using about 5 ml of N/2 NaOH. Dilute to 1 litre. Standardise against standard Ca hardness solution, which should generally gives 1 ml EDTA equivalent to 1 mg $CaCO_3$.This solution is used for water samples having high hardness content. For low hardness content water, dilute the above solution ten times to get 1 ml EDTA equivalent to 0.1 mg $CaCO_3$.

Hardness Indicator: 0.5 g Na_2CO_3 + 10 g eriochrome black T. The mixture should pass through 20 mesh. Take 0.45 gm of this mixture per 60 ml acidified methanol (0.5 ml. conc. HCl per 1000 ml. methanol). Hardness indicator gives red colour when complexed with calcium & magnesium. With EDTA addition, it complexes calcium & magnesium and finally the indicator regains a free form to turn the sample blue.

Buffer Solution: 90 g NH_4Cl + 560 ml of sp gr 0.9 NH_4OH. Dilute to 950 ml with pure water. On dissolution, add 0.83 g sodium salt of EDTA + 0.5 g $MgSO_4.7H_2O$ and dilute to 1 litre. Standardise this by taking 2 ml of this buffer in 50 ml pure water with two drops of indicator. One drop of EDTA should change red to blue and one drop calcium solution (1 ml equivalent to 0.855mg $CaCO_3$) should change blue to red.

For testing of high hardness water sample, take an aliquot of 50 ml. Add to it 1 ml of buffer + 4 drops of indicator and titrate with high hardness EDTA solution. Reading multiplied with 20 gives ppm of total hardness as $CaCO_3$. For low hardness sample of water, take 100 ml of aliquot. Add to it five drops of buffer and 4 drops of indicator. The titre reading itself gives ppm of total hardness as $CaCO_3$. For getting sharp end point with a very low hardness water, it is better to use one drop of hardness sulphide solution (prepared by dissolving 5 g of NaOH + 5 g of $Na_2S.9\,H_2O$ and diluted to 100 ml with pure water). For calculating total hardness.

Between total hardness (TH) & total alkalinity (TA), the following relations can be obtained:

When TH > TA	Carbonate (temporary) hardness = Total alkalinity
When TH < TA	Carbonate (temporary) hardness = Total hardness
When TH = TA	

Non-carbonate (permanent) hardness = Total hardness – Total alkalinity. Magnesium hardness is mostly calculated by substracting Ca hardness (separately determined by test and assuming contribution from other metallic ions are in traces).

Magnesium as Mg, mg/l, then	0.2430 (TH as $CaCO_3$ – Ca as $CaCO_3$)
If TA > TH, Alkali metals as Na mg/l	0.4596 (TA-TH as $CaCO_3$) + 0.4789 (SO_4 as SO_4 mg/l) + 0.6486 (Cl as Cl mg/l)

If TA = TH, alkali metals will be calculated from the amounts of SO_4 using factor as above. Excess alkalinity is the alkalinity in terms of Na_2CO_3 equal to the excess of total alkalinity over total hardness i.e. = 1.06 (TA –TH)

Calcium Hardness:

The test is based on the titration of 50 ml sample with same EDTA solution as above under total hardness testing, but at pH above 12 (made by adding 2 ml. of 1N NaOH solution to above aliquot) and added with 0.2 g approx. indicator (prepared by grinding 0.45 g of murexide (ammon perpurate) with 10 g of NaCl to a fine powder) or by adding 5 drops of indicator solution (by dissolving 0.45 g of murexide in 50 ml triethanolamine). Magnesium are precipitated as magnesium hydroxide and does not interfere in test, but the titration should be performed immediatly after addition of NaOH. The indicator changes from a salmon pink to an orchid purple at the end point. The calculation of calcium hardness as $CaCO_3$ mg/l is made on the basis of strength of EDTA solution as above.

Orthophosphate, if present, in water will however precipitate calcium at the pH of the test. Sr & Ba interfere with the calcium determination and alkalinity in excess of 30 mg/l may cause an indistinct end point with hard water. No other element causing hardness like iron, aluminium, zinc, lead, tin, etc. will interfere with the test if found less than 5 mg/l.

Chloride:

In the titrimetric method for chloride, a standard solution of AgNO3 (N/71 i. e 1 ml equivalent to 0.5 mg of Cl^-) is used as titrant and potassium chromate (10% solution - 5 drops) as indicator. The chloride ion sample reacts with the silver ions in the titrant to form insoluble silver chloride. When all of the chloride has been precipitated out of solution as silver chloride, the excess silver ions react with the chromate ions to form insoluble silver chromate, which is a brick red colour.

For a 100 ml water sample, the calculation for the chloride content mg/l as Cl is done like this: (Titre - 0.4) x 5, where 0.4 is taken as blank for a 100 ml. chloride free sample, which is accounted for the formation of $AgCr_2O_4$ by whose colour, the end point is detected & thus more $AgNO_3$ is consumed. If the colour is deep, then more blank is to be accounted, but for just faint colour, 0.4 ml. is sufficient.

Notes:

a. It is better to neutralise the water sample with N/2 H_2SO_4 using phenolphthalein indicator before titration, since chromate indicator works best in neutral medium. Otherwise, the titration can be carried out in the sample already undertaken for phenolphthalein alkalinity test.

b. Influence of sulphite content greater than 10 mg/l, the sulphite content should be oxidised by 4 ml of 3% hydrogen peroxide to prevent interference with the chloride test.

c. Nitric acid, add 8 ml of nitrobenzene followed with an excess known amount of $AgNO_3$ (say 20 ml) and back titrated with standard ammonium thiocyanate solution using ferric ammon sulphate as indicator.

Sulphite:

The titration of sulphite is based on the reaction of sulphite with iodine in acidic solution. To a 100 ml. water sample add 5 ml HCl (1:3) plus 2 ml starch and 5 ml KI (3%), and titrate with N/63 KIO appears.

Notes:

The water sample should be freshly obtained with a little contact with air as possible & hence do not filter, but cool to 20 - 25^{0}C. The method is also affected by any oxidisable substances present in water like organic matter, sulphides and nitrites giving high results.

Carbon Dioxide:

Free carbon dioxide reacts with sodium carbonate to form sodium bicarbonate. At the equivalence point (pH 8.3), the phenolphthalein indicator turns pink. If free mineral acidity is present (viz., HCl or H_2SO_4 as in case of decationised water in DM plant), it is measured along with the free carbon dioxide. To correct this, a second sample is titrated with the sodium carbonate to a pH of 4.3. Substraction of the two titrations provides a corrected free carbon dioxide concentration.

The sample is obtained by running the water through a polythene tube discharging into the bottom of a 100 ml. graduated cylinder and withdraw the tube while the sample is still overflowing. By a suitable pipetting arrangement excess sample above the 100 ml mark is then withdrawn followed by immediate addition of drops of phenolphthalein indicator. Absence of any pink colour indicates absence of carbon dioxide or free mineral acidity, but if the sample remains colourless, it is titrated with N/22 sodium carbonate solution to pH 8 until pink colour stabilizes.

Bicarbonate is determined by stripping carbon dioxide from an acidified sample (10 litre) by means of a purified air stream through a suitably made apparatus and absorbed in a known amount of standard NaOH solution. The quantity of gas is calculated from the amount of acid needed to titrate the hydroxide solution from pH 8.3 to pH 4.3, using the formula.

$$1mL\ N/100\ H_2SO_4 = 440\ ppb\ CO_2$$

The free CO_2 content in natural water can also be approximately assessed by the below graph showing relationship between alkalinity, pH and free CO_2. However, this graph applies at ambient temperatures to natural surface water, which do not contain sodium bicarbonate, i.e. the graph applies only when alkalinity < total hardness.

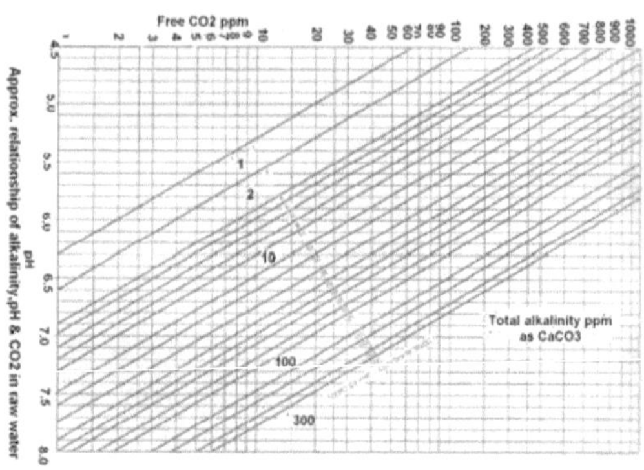

Total organic matter: Organic compounds are generally measured as permanganate demand ($KMnO_4$) of water or chemical oxygen demand (COD) or biochemical oxygen demand (BOD). Micro/macroorganisms in water use organic matters as a food source, for which a biochemical reaction exists, in which dissolved oxygen of water is consumed as the end product of water and carbon dioxide is formed, causing demand of oxygen by water, proportionate to the organic matter present in it.

KMnO4: The demand of oxygen by this method is measured by oxidising the organic matters under both alkaline and acidic conditions, by taking 100 ml water sample followed with first 5 ml 15% NaOH and 15 ml. N/100 $KMnO_4$. Immediately heat to boil for 16 minutes & add cautiously 7 ml of 1:1 H SO_4. Continue boiling for further 30 minutes. Add 15 ml. of N/100 oxalic acid. Wait for few seconds & titrate with 0.01N $KMnO_4$.

$KMnO_4$ consumption = Oxalic acid consumption x 3.16 mg/l

(Reading to be reduced by 0.6 - 0.8 ml. for blank, correcting for auto decomposition of $KMnO_4$ under the conditions of the method). However, for general natural water, which is not contaminated considerably with organic matters, it is general practice to oxidise the humic and fulvic acids normally present in natural water in acidic medium only, by taking 100 ml water sample along with 10 ml. 25% sulphuric acid and 10 ml N/80 potassium permanganate solution and oxidise either by keeping the mixture exactly at 37 ^{0}C for 4 hours or boil the same at 100 ^{0}C for 30 minutes. After this 10 ml of N/80 oxalic acids are added to it, and back titration is conducted to know the exact amount of $KMnO_4$ consumed. The total potassium permanganate consumption of the sample should not exceed 5.0 ml after deducting the blank reading taken under exactly same fashion. If it does, dilute the sample appropriately with pure water and repeat and calculate as:

1 mL N/80 $KMnO_4$ = 1 mg/L as oxygen consumed and expressed as $KMnO_4$, mg/L O_2 consumed at 100 deg C, which when multiplied with 0.5 gives the value of $KMnO_4$ as mg of oxygen consumed (4 hours; 37 deg C).

NOTES:

Interferences due to inorganic reducing agent, present if any, needs to be accounted by titrating a separate same aliquot of water sample mixed with 10 ml. 25% sulphuric acid with same N/80 ml. potassium permanganate till a permanent pink end point for 3 minutes, and the consumption value (minus the blank) is deducted from the ml. consumed in above main test for calculating out the actual KMnO4.

COD: This is determined by refluxing a water sample with chemicals as below in the proportion as shown and titrated: 5 ml sample of water mixed with a cold solution of 5 ml. N/8 $K_2Cr_2O_7$ and 10 ml conc. H_2SO_4 and 1 ml of saturated $AgSO_4$ (50%). The mixture is then refluxed for 2 hours avoiding bumping. The reflux is then cooled and 45 ml. of water along with one drop of ferrous-phenonthroline indicator (a solution of 3.475 g $FeSO_4.7H_2O$ in 500 ml water well mixed with 7.427 g 1:10 phenonthroline). Titrate the residual dichromate with N/8 $FeSO_4$ solution (34.75g/l $FeSO_4.7H_2O$) and oxygen consumed is calculated as: 1 ml N/8 $K_2Cr_2O_7$ = 1 ml oxygen

NOTES:

Overheating should be avoided as there may be breakdown of dichromate to chromic salts. Substracting the sample titration from the blank titration gives the volume of potassium dichromate equivalent to the oxygen demand.

$$COD = \frac{(\text{Blank titration} - \text{Sample titration})}{\text{Volume of sample in mL}} \times 1000\ mg/L$$

Also, this procedure gives obscure results in case of presence of more than 100 ppm of chloride in the sample. In case of more chloride, addition of more $AgSO_4$ is needed.

BOD: BOD is actually a measure of the oxygen consumed by microorganisms as they assimilate organics and is determined by measuring the loss in dissolved oxygen in water sample after incubating it for 5 days at 20 ^{0}C.

The sample of water is first dissipitated with available chlorine after standing 1 - 2 hours or neutralised with sodium sulphite. If deficient in oxygen, the sample is aerated to bring oxygen concentration up to but not in excess of saturation at 20 ^{0}C. A definite volume of sample is taken in each of the two BOD bottles and filled completely with 'dilution water' consisting of phosphate buffer, $MgSO_4$, $CaCl_2$ and $FeCl_3$ and seeded with by adding 1 drop of lactose broth and further suitably aerated with a clean supply of compressed air to saturate with oxygen. Saturation of the 'dilution water' may also be obtained by 2 - 3 days' storage in partially filled bottles closed with cotton plugs. Determine dissolved oxygen in one bottle before incubation, and in the second bottle after an incubation period of 5 days at 20 ^{0}C (water baths with temperature controlled to 20 +/- 1 ^{0}C may be used as incubator) and loss of dissolved oxygen estimated.

Dissolved Oxygen:

The dissolved oxygen in ppm level is generally measured by Winkler procedure entailing the oxidation of manganous hydroxide in a highly alkaline solution. Upon acidification in the presence of an iodide, the manganic hydroxide dissolves and free iodine is liberated in an amount equivalent to the oxygen originally dissolved in the sample. The free iodine is titrated with a sodium thiosulphate standard solution using starch as an internal indicator, after most of the iodine has been reduced. The normality of the thiosulphate solution is adjusted, so that 1 ml = 1mg/l dissolved oxygen when 200 ml of the original sample is titrated.

NOTES:

1. When the BOD value of any sample is expected to be in excess of 5.0 mg/l, dilution of the sample will be necessary.
2. Nitrites, ferrous [above 1 mg/l & ferric iron (on delayed titration)], organic matters, sulphides, sulphites, polythionates, hypochlorites, suspended matter and other oxidising & reducing substances interfere with this test procedure either by absorbing or reducing the liberated iodine or by oxidising iodide to free iodine.
3. Testing for BOD is not usually undertaken under thermal power station chemistry control practices, but its value is important for designing for efficient water treatment plant operation, since this gives the following relations:
    - ppm chlorine dose is generally twice the BOD value
    - Ratio of KMnO (as O)/BOD is generally 1:3 or 1:4 in sewage contaminated water
    - In domestic untreated waste, BOD/COD = 0.4 to 0.8 and BOD/TOC= 1.0 to 1.6

Total organic nitrogen:

This test is not usually undertaken in a TPS chemical laboratory, but has relevance in substantiating the organic matter present in considerable amount. This is generally food for microorganisms for its growth & multiplicity, and in the process the decomposed products like ammonical nitrogen, nitrites and finally nitrates are generated. Large ammonical nitrogen indicates recent contamination of decomposed products of organic matter. Ammonia is also favourable for development of certain bacteria, which in turn, gives rise to unpleasant taste.

Before undertaking this test, it is necessary to remove free ammonical nitrogen by distillation at pH near 7.4, preferably by using phosphate buffer. The ammonia concentration in the distillate can also be measured as ammonical nitrogen by making same to a fixed volume by dilution, and titrating its suitable aliquot with standard acid or ammonia content seen colourimetrically by Nesseler reagent method (described under colourimetric testing method). Then the sample is digested in alkaline medium (preferably alkaline potassium permanganate (8 g $KMnO_4$ +200 g NaOH per litre solution, in presence of H_2SO_4 & $CuSO_4$ as catalyst. Amino nitrogen of many organic compounds is converted to ammonium bisulphate, which in alkaline medium liberate ammonia. The distilled ammonia is subsequently absorbed in known amount of sulphuric acid; the excess acid is then back titrated to know the amount of acid consumed and the total organic nitrogen present in the form of the liberated ammonia.

Nitrite along with nitrates, if any present, if required for measurement, as nitrite/nitrate, nitrogen can be reduced to ammonia by distilling with NaOH solution (250 g per litre, suitably treated with 2 or 3 strips of aluminium foil for atleast 12 hours) and ammonia content measured as stated above. By substracting the nitrate nitrogen amount as determined under the colourometric test procedure described later, the nitrite nitrogen alone can also be found.

Free NaOH (in boiler water):

This is determined by the barker's equation as follows:

$$\text{Free NaOH, ppm} = 4f\,(\text{ml N/100 acid}) - 0.8\,(1.05\ PO_4\ \text{ppm} + 1.66\ SiO_2\ \text{ppm})$$

However, under normal operating conditions, ammonia presence allowance has also to be given or eliminated by passing the sample through a cation resin column.

$$\text{Allowance x ppm } NH_3 = 2.35 \text{ x ppm NaOH}$$

Instrumental Methods Direct Read Out

PH: pH is a measure of the active acid or base in a solution and is defined as the negative logarithm of the hydrogen ion concentration. A neutral water has a pH value of 7. A value below 7 & approaching 0 are increasingly acid while values above are increasingly alkaline. The determination of pH value enables water to be divided into distinct classes:

- pH>8: contains no free carbonic acid but contains carbonates, with or without bicarbonates
- pH betweenH 4.5 & 8: This group includes great majority of water; contains no carbonates but contain carbonic acid & bicarbonates. If the pH exceeds 4.5 and after boiling to drive off the free and semi-combined carbonic acid, the value exceeds 7, carbonic acid is the only acid present
- pH < 4.5 & 7 remaining under after boiling: Contain some free acid besides carbonic pH is conveniently measured with a pH metre (electronic & read out devices). Although the hydrogen gas electrode is recognised as the primary standard, the glass electrode in combination with the reference potential by a saturated calomel electrode is most generally used. The glass electrode system is based on the fact that change of 1.0 pH unit produces an electrical charge of 59.1 millivolts at 25^{0}C. The error caused by high sodium ion concentration at a pH above 10 may be reduced by using special 'low sodium error' electrodes or by correcting from the chart supplied by the manufacturer. Similarly for temperature, correction is made with the chart, in case the supplied instrument does not have inbuilt temperature compensation device. The temperature exerts two significant effects on pH measurement: (a) the electrode themselves vary in potential with temperature (b) ionisation in the sample varies with temperature. Rest operating procedures have to be followed from the instruments manual.

Standardisation of The Instrument:

The standardisation of the electrode and the instrument is usually done with three different buffer solutions:

pH 4	Anhydrous potassium biphthalate (10.2 g/l)
pH 7	Anhydrous potassium dihydrogen phosphate (1.361 g) + anhydrous disodium hydrogen phosphate (1.420 g) per litre solution
pH 9	Sodium tetraborate deca hydrate (3.81 g/l)

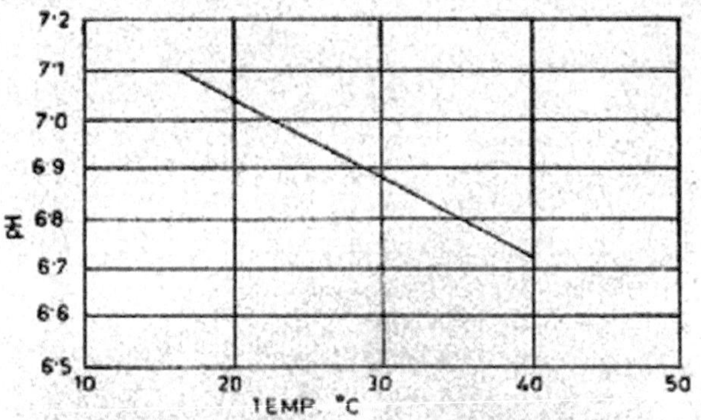

EFFECT OF TEMP. ON PH OF PURE H_2O

Conductivity:

Ionisable solids in water give it the ability to conduct electrical current and if this current is measured by a suitable conductivity metre, the purity of water with respect to dissolved salts can be assessed, in view of the fact that at infinite dilution, all dissolved salts in natural water are supposed to be fully ionised. Thus, lesser the conductivity, purer is the water. Conductivity scales in the metre are calibrated as micromhos/cm or microsiemens/cm and the conductivity is measured using a conductivity cell having two platinum electrodes, and which are of different cell constants ranging from 0.01 to generally 1.2 with selective use to very pure water to a water having very high dissolved salts. The current generated by the conductivity cell passes through the winding of the ratio transformer, the output of which is further amplified to read the conductivity reading as specific conductivity of the sample water.

For the rest procedure of tests, operating instructions from the instruments manual are needed to be followed. If temperature compensation device is not built in the instrument, the correction is needed to be made from the graph, to express the conductivity figure always at 25^{0}C.

Standardisation of the Instrutment: Calibration of the cell and the instrument is done with standard solution, say 0.1 N KCl (aqueous) which has specific conductivity of the order of 12.88 micromhos/cm at around 25^{0}C and 11.67 at 20^{0}C. For 0.01N KCl these are 1.413 and 1.278 micromhos/cm respectively.

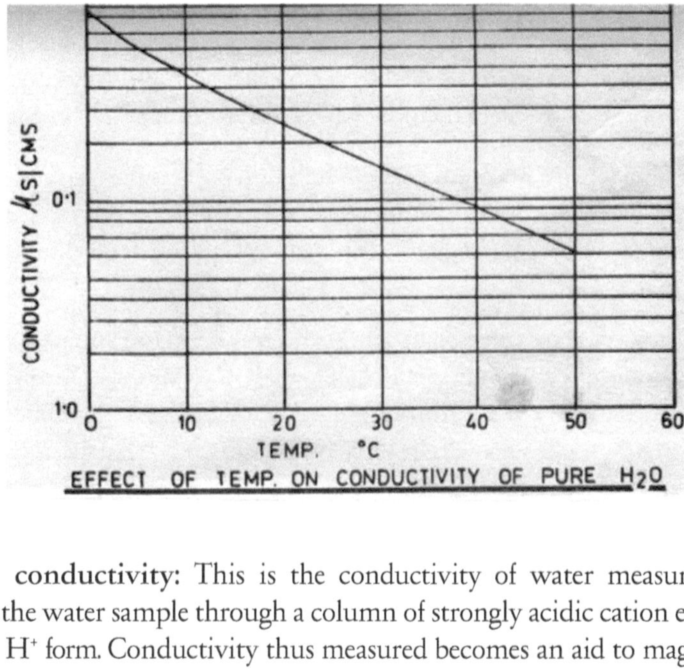

EFFECT OF TEMP. ON CONDUCTIVITY OF PURE H_2O

Cation conductivity: This is the conductivity of water measured after passing the water sample through a column of strongly acidic cation exchange resin in H^+ form. Conductivity thus measured becomes an aid to magnify the ionised salt impurities of anions like chloride, sulphate, etc. present in water in submicro level range, and will show an enhanced conductivity due to the conversion to corresponding acidic equivalent. Under TPS practice, this is particularly very useful in detecting even minor condenser leak. The condenser leak will introduce chloride ions in the form of NaCl, $CaCl_2$, etc. ions, the conductivity of which may not be significant in view of the contamination at very low level, but when the conductivity is measured after passing through the resin column, the exchange reaction of the resin will produce hydrochloric acid, and the conductivity then will increase manifold giving clear indication of impurities contamination.

$RH^+ + NaCl^- \rightarrow RNa^+ + HCl$, where R represents the resin matrix.

Similarly, for boiler feed water & steam, the contamination of impurities is easier to know, if the contaminations are anionic in nature.

Sulphate: It can be estimated by substituting the amount of chloride in ppm as $CaCO_3$ by another method. Under TPS practice, boiler feed water & steam cycle sample water should not contain anything else than ammonia, which after removal should indicate the purest form of water having conductivity 0.1 to 0.2 miromhos/cm conductivity. The impurities of other cations, if any, including sodium however remain undetected by the method of measuring conductivity. These are measured by other methods particularly with specific ion analysers.

Partially exhausted cation column will however deliver neutral effluent, rather than a true acid effluent, in which case, cation conductivity will be erroneously low. On final exhaust, conductivity will be high due to leakage of ammonia.

Cation exchange column height shall be 90 cm and internal diametre 25 mm. The resin bed height shall be 50 cm. Top & bottom of column shall be equipped with nylon chord pad. It shall be complete with rubber tubing, stop cocks, 6 mm glass delivery tube & other accessories. Smallest dimension of lab column could be with internal diametre 1.5 cm, height 20 cms & rate of passing 10 ml/min.

Turbidity:

Turbidity is generally measured by nephlometre, which works on the principle of scattering of light by the suspended matters present in water responsible to cause turbidity in water. When a strong light beam is passed upward through a tube (called sample cell) containing the sample of water to sense the scattered light at 90^0 to the incident light by the photo tube, it gives turbidity reading in NTU (nephlometric turbidity unit). Rest testing procedures are followed as per the operation manual of the turbidity instrument. Another turbidity units as used are FTU (formazin turbidity units), measured on the principle of measuring the reduction of light intensity caused by scattering and absorbtion as light passes through the sample, by an electro photometre. However, this FTU is found equivalent to NTU, when readings are made on nephlometre.

Standardisation of the Instrument: The instrument is standardised by 'formazin' suspension, made by mixing 5 gms of reagent grade 'hydrazine sulphate' per 400 ml pure water and 50 gms of pure 'hexamethelene tetramine' per 400 ml same water properly and then made up to 1 litre suspension by the same water & mixed. This mixture is then allowed to settle for 48 hours (22 to 28 ^{0}C) during which, it will develop suspension particles of uniform size & shape. This suspension of 4000 NTU/FTU is treated as stock suspension and is stable for 6 to 8 months. Before preparing further standard dilute solutions, the stock is needed to be shaken thoroughly and this diluted suspension is required to be prepared on daily basis, giving standard solution within 1% accuracy when properly mixed.

NOTE:

The amount of turbidity registered is dependent on such variables as the size, shape and refractive properties of the particles. Hence, no direct relationship exists between the turbidity of water sample and the weight concentration of the suspended matter present.

Instrumental Methods

Colourimetric Method

Residual Chlorine:

Lovibond comparator is the most comprehensive laboratory system available for routine chemical analysis for this. This consists of two compartments, each housing similar transparent cells to contain water sample (one for test and the other for blank) to certain optical depth. The blank sample compartment also houses standard test discs, with some of the comparator housing having lighting system from beneath the cells containing sample and blank.

After taking the requisite amount of sample and blank in each of the cells, few drops of acid ortho-toluidine (pH below 2 by HCl) solution proportionate to the optical depth of water sample is added in the sample cell and the yellow colour developed matched with the disc colour by rotating the disc in front of the cell having same water sample as blank. Use of DPD (N, N-diethyl-p-phenylenediamine) is also made in place of ortho-toluidine, but the red colour produced by this is matched with separate standard discs. This test is called 'Palin test', and in some country, toluidine test is withdrawn in view of ortho toluidine being recognised as capable of causing cancer.

Yellow colour as produced by o-toluidine could also be due to presence of nitrite or manganese and could be due to combined chlorine as present in the water. On adding a very small crystal of sodium thiosulphate, the yellow colour due to presence of free chlorine or combined chlorine (total residual chlorine) will be discharged, but if there are any nitrites or manganese, the solution will remain yellow. On substracting of this colour as per disc, from the original colour, gives the true value of chlorine.

For determining free and combined chlorine (chloramines, etc) separately, testing needs to be conducted by taking 3 separate sample cells as below:

1. Contain o-toluidine reagent + water. Mix quickly & immediately. Compare the colour within 5 seconds from standard colour. Add arsenite reagent & mix quickly again and compare with standard colour as rapidly as possible. This value (A) represents free available chlorine & interfering colours.
2. Contain arsenite reagent + water. Mix quickly & immediately add 0-toluidine reagent. Mix quickly again and compare standard colour as rapidly as possible. Record the result as (B_1) value. After exactly 5 minutes, compare again with standard colours & record this as (B_2). This represents the interfering colours immediately and after 5 minutes respectively.
3. Contain 0-toluidine reagent + water. Mix quickly & after exactly five minutes, compare the colour with standard and record the value as OT. This value represents the total amount of residual chlorine and total amount of interfering colours.

Thus, total residual chlorine	$OT - B_2$
Free available chlorine	$A - B_1$
Combined available chlorine	Total residual chlorine – Free available chlorine

NOTES:

a. If water is alkaline, double the dose of toluidine reagent or neutalise water with HCl.

b. The standard colour can also be prepared by mixing chromate & dichromate solutions in requisite proportions. The solutions, however, need to be in preserved air tight containers.

c. DPD reagents are generally supplied as tablets with detailed instructions. Determination of combined residual chlorine with this, depends on its content of KI, for which necessary reagents can be prepared as solution.

d. Lovibond colour comparator, other water contents can also be analysed by developing colour with suitable reagents and comparing the colour with standard discs specifically meant for it. For this, a modified form of comparator commonly known as nessleriser (or aqua tester) can also be used. These colourometric methods, however, have become popular because of their simplicity and relatively low cost, but are not accurate as the photometric/ spectrophotometric methods.

Dissolved Oxygen:

Removal of dissolved oxygen from boiler feed water is very important because of the accelerated effect oxygen has at the boiler operating temperatures & pressures in respect of corrosion in the entire system. Hence, it is necessary to measure/detect even very low level of oxygen (ppb range) typically associated with boiler operation.

The basis of test is that the reduced alkaline solution of indigo-carmine dye reacts even with traces of dissolved oxygen to give increasing intensity of colour ranging from orange tinge to deep purple proportionate to the dissolved oxygen content while in its reduced form, the dye is having bright yellow-green colour, meaning thereby no oxygen, as below: The indigo-carmine is reduced by glucose in the presence of potassium hydroxide, and glycerol is used to provide air stability and to sharpen the colours.

Colour	Oxygen content, ppb
Yellow	0
Orange	5
Orange pink	10
Pink	15
Pink red	20
Red purple	50
Purple	100

The test is however needed to be conducted in flowing stream of the sample water to avoid atmospheric oxygen contamination, and for the purpose specially designed narrow mouth bottles with air tight stopper (called DO bottle) are taken including those of having standard colours as above hot samples are beforehand cooled online with the flowing one.

The cooled flowing sample (the temperature must not exceed 21 ^{0}C) is run to bottom of the DO bottle (300 ml) through a glass tube well connected from the flowing tube through a suitable rubber tube and allowed to overflow the DO bottle for few minutes. Then insert a 5.0 ml pipette having indigo-carmine dye solution up to mark into the bottle up to bottom without releasing the dye solution. Then while releasing the dye from the pipette with one hand, stopper the bottle with other hand immediately after removing the DO bottle from the flowing stream. Mix the contents of the bottle well and then compare the colour from standard solutions kept in similar DO bottles.

Sulphite and hydrazine are known not to interfere in the test, but organic oxygen scavengers, if used in the system, may give positive interferences. Similarly no interference is noted in presence of Ni, Cu or Zinc in concentration of 1 mg/l or from ferric ion at 3 mg/l, but ferrous ions are known to have interference, and hence the same is removed by a mixed bed ion exchange demineraliser in series with the sampling line.

Indigo-carmine dye solution (called Leuco solution) is prepared by mixing 20 ml. of indigo-carmine glucose solution (0.018g indigo carmine plus 0.2 gm of dextrose dissolved in 5 ml of distilled water and mixed well with 75 ml. of glycerol and kept in dark place to be used before two weeks expiry) and 5 ml of KOH solution (50 g/100 ml). After thorough mixing, allow the reagent to stand for 10 minutes until the initial dark red colour changes to lemon yellow. The reagent thus prepared, should be consumed on the same day of its preparation. The colour standards to compare with the sample are beforehand kept prepared in similar type 300 ml. DO bottles by mixing red, yellow and blue colour standards as below:

Dissolved oxygen ppm	Red colour standard (59.29gm cobaltous hexahydrate in 1:99 HCl to make 1 litre)	Yellow colour standard (45.05gm ferric chloride hexahydrate in 1:99 HCl to make 1 litre)	Blue colour standard (62.45 gm cupric sulphate pentahydrate in 1:99 HCl to make 1 litre
	"A" in ml	"B" in ml	"C" in ml
0	0.75	35.0	--
0.005	5.0	20.0	--
0.010	6.25	12.5	--
0.015	9.40	10.0	--
0.025	14.40	3.8	--
0.044	16.02	2.0	2.76
0.062	23.0	1.0	18.18

To each of the 300 ml DO bottle, 2.3 ml of HCl is added and diluted to neck of the bottle with water and stoppered to mix further by inversion. All these bottles are then stored in a dark place to minimise the fading of colours.

Aluminium:

Determination of aluminium in water is generally required to check clarifier performance under pretreatment of water, where to remove turbidity alum or aluminium compounds are used and an excess residual of which in water is harmful.

Summary of method is that haematoxylin reacts with aluminium ions in water to give purple colour at pH 7.5 +/- 0.2 and the colour is proportional to the concentration of aluminium. Rich golden colour will mark the absence of aluminium. Range of application is up to 0.5 ppm aluminium as Al. Calcium above 100 ppm and magnesium above 250 ppm are likely to interfere. Manganese above 0.2 ppm causes bleaching and iron above 0.4 ppm gives shades usually of greenish pattern. In alkaline solution haemotoxylin gives violet colour lake with aluminium, while in acid medium, the lake is stabilised & the colour of excess dye changes from red to yellow.

To a 50 ml sample, 1 ml of saturated ammonium carbonate (15%) and 1 ml of 0.1% haemotoxylin added and the well mixed mixture is allowed to stand for 15 minutes. Then add 2 ml of 30% acetic acid to it and mix properly and compare the colour with standard colour disc of aluminium rotating in front of the blank solution tube cell of the colour comparator, the blank with pure water also added with same chemicals step by step.

If in the sample, some precipitated aluminium is also suspected, then it should be dissolved by adding few drops of 5N HCl and made neutral with the help of ammonium carbonate before proceeding for test.

Nitrate Nitrogen:

The sample is treated with sodium salicylate (5 g dissolved in 50 ml 1N solution and further diluted to 1 litre) in presence of conc. sulphuric acid; the mixture is then made alkaline with NaOH solution (4N) and the colour obtained is matched against that produced with a series of standard nitrate solutions.

The volume of the sample is so choosen by dilution or evaporation that 25 ml shall contain not more than 0.02 mg of nitrates as N. This is transferred to a porcelain dish and 5 drops of NaCl solution (2%) and 1 ml of sodium salicylate added to it is evaporated to dryness on a water bath. Then the residue is moistened after cooling with 1 ml. conc. sulphuric acid and allowed to react for exactly 10 minutes. The contents of the dish is then transferred quantitatively to a nessler tube and 10 ml of NaOH solution added to it and made up the volume up to mark with water. The colour so obtained is then compared with a series of standard nitrates solution treated similarly and amount of nitrate nitrogen is known.

Instrumental Methods

Photometric / Spectrophotometric

Photometres or spectrophotometres provide the most accurate means of measuring the colour of a reacted sample. These methods are based on the principle of Lambert-Beer law, which states that when a monochromatic light beam is passed through a dilute solution of light absorbing compound, the amount of light emerging from the solution is normally governed by the equation:

$Log_{10} I/I_0 = e.cxl$; where	I =	Intensity of emergent beam of light
	I =0	Intensity of Incident beam of light
	e =	Molar extinction coefficient (a constant) of light
	x	absorbing compound at wave length x
	c =	Molar concentration of light absorbing compound
	l =	Length of solution (cms) through which light passes

Thus, on obey of the law, a graph of absorbance against concentration is a straight line and if absorbance is measured by an instrument, in such a way that absorption due to the solvent & container (cuvette cell) used to hold the solution is canceled out; the graph should pass the origin.

Hence, for a given length of solution, when e is known, measurement of I & I photoelectrically allows an estimate of 'c' to be made. Absorbances are measured either by spectrophotometre or absorptiometres. In spectrophotometre, the wavelength of the incident light beam may be continuously selected by a monochromator, where as in absorptiometres (simple photometre), the wave length of the incident light beam is selected by means of suitable coloured filters.

In spectrophotometre, the tungsten lamp is used for the wavelength 320 - 1000 nm while Deuterium lamp is used for 200 - 320 nm. Two types of photo tubes (a) ultra violet sensitive photo multiplier (b) Infra red sensitive photo tube. Photo multiplier is used for ultraviolet & visible range due to high resolution of the grating (obtained by superior silt width), a narrow band width is obtained.

Thus, whatever method is adopted to measure the light absorption from the reacted colour sample, the sensitivity of a reagent matters most, which is defined as the number of micrograms of determined per cm^2 cross sectional area of the light absorbing solution, that are required to give an absorbance of 0.001 at a specified wavelength. Based on the concept, laboratories in general now use the spectophotometres for analysis involving light absorbance by the sample as below:

Silica

Silica is determined in the laboratory usually by developing colour of molybdenum blue complex and measuring its intensity. Concentration of silica in excess of 25 ppm in natural water frequently contained dimers and trimers. Molybdenum blue complex detection method of silica works up to the dimers only (it is called reactive silica). Higher polymers, viz., trimers and above with other low molecular weight polymers, and range on up to quite large particles easily filtered by absolute filters (called colloidal silica) go undetected by molybdenum blue complex method. Thus, in any water silica could be present in two forms:

1. Reactive
2. Colloidal

In thermal power applications, chances of presence of colloidal silica in boiler feed water steam cycle are remote in view of system operating alkalinity & temperature, at which colloidal silica, if any, present, in make-up water (demineralised water) becomes reactive. From raw water to demineralised water, the presense of colloidal silica also, cannot be ruled out. Hence, determination of both forms of silica under TPS practice is found necessary.

Reactive Silica:

Silica and phosphate in dissolved state in water sample react with molybdate ion under acid conditions to form a yellow colour due to the silico molybdic and phospho molybdic acid complexes. Citric/oxalic/tartaric acid are added, which mask the phospho molybdate acid complex and only the silico molybdic acid complex remains for test. If the dissolved silica are present in high amount (cf. raw water etc), the yellow colour of silicomolybdic acid complex is intense enough to be measured directly by spectrophotometre at 450 nm wavelength, or the high concentration is made low (preferably below 1 mg/l) by suitable dilution. Low concentration of molybdosilicic acid giving faint yellow colour is then treated with an amino acid reducing agent (1, amino, 2, napthol 4, sulphonic acid), which give dark blue colour of molybdenum blue complex and measured by spectrophotometre at 815 nm wavelength. The amount of colour formed in either case is directly proportional to the amount of silica present in the sample and is measured by the predetermined calibration curve. The curve is however found linear to atleast 0.5 mg/l. A good sensitive spectrophotometre will generally give 0.13 absorbance of the blue colour of 0.1 mg/l silica after deducting suitably against the blank reading. Colour and turbidity interferences are eliminated when the instrument is zeroed with the blank and the reagent blank reading is separately recorded. A 40 mm cuvette cell for holding the sample is also essential for determination of low amount of silica.

For low level determination of silica, to a 100 ml water sample, 2.5 ml of ammonium molybdate + sulphuric acid reagent (a solution of 75 g ammon. mollybdate in 500 ml of silica free water, mixed with 322 ml of 10 N H_2SO_4 and finally made up to 1litre with silica free water) is added and then a reaction time of five minutes is allowed. After that 2.5 ml of tartaric acid (10% w/v) solution is added and again waited for 5 minutes after quick mixing. Finally, 2 ml of ANSA reagent (10% solution of amino mixture prepared by mixing 5 g of sodium sulphite & 2.75 g of 1 amino 2 napthol 4 sulphonic acid in a mortar to a fine powder and then mix again with 140 g of sodium meta bi sulphite to pass through 20 mesh) is added and waited for 15 minutes to form the stable colour. Intensity of this colour is then measured at 815 nm light wave length by recording the absorption by taking the blue solution in a 40 mm cuvette cell and noting the silica content from the calibration curve prepared in the same fashion including deduction of the blank reading.

Colloidal Silica

Colloidal silica is the undissolved part of the total silica present usually in raw water. Hence, after determination of reactive silica in the sample, a separate sample filtered through 1.2 micron filter (to remove finely divided sand particles) is taken and total silica is determined on it after dissolution of all the colloidal silica present as below, in the same manner as determined against reactive silica as above. Then colloidal silica content comes out to be as:

Colloidal silica = Total silica – Reactive silica

The dissolution of the colloidal silica is then undertaken by the following methods:

1. 100 mL of filtered sample is taken in a platinum basin and heated to dryness slowly in a dust free environment (large aliquot has to be taken if colloidal silica is expected less than 1 mg/l). After dryness, heat the basin to redness slowly and then cool. Mix about 0.2 to 0.5 g of pure anhydrous sodium carbonate and heat slowly to melt first and then to bright redness of the crucible to maintain like this for further 10 minutes. Allow the crucible to cool and dissolve the melt in silica free water and extract the same into a volumetric flask to make it finally 100 ml mark to test for reactive silica in the fashion as described above.
2. To a 100 ml of filtered sample in a polythene container, 4 ml of mixed acid reagent [by mixing solution 'a' (20 ml 42 - 43% HF per 400 ml water, with solution 'B' (28 ml 98% sulphuric acid per 400 ml water and made up to 1 litre in a polythene volumetric flask and stored at 70 ^{0}C for 24 hours after due mixing] is added and kept in water bath at 70 ^{0}C for 16 hours. Then to it 8 ml. of aluminium reagent [prepared by dissolving 227 g of aluminium ammonium sulphate i.e. 0.25M of (NH_3). $Al_2(SO_4)_3$. 24 H_2O in 900 ml of water and made up to 1 litre after adding 1 ml of 98% sulphuric acid and stored at 50 - 70 ^{0}C to prevent crystallisation at room temperature, the solution is supersaturated] is added and kept at same water bath of 70 ^{0}C for 45 minutes & then cooled. After that usual determination method or reactive silica as laid down above is followed.
3. When colloidal silica is present at low levels (< 1 ppm), it is also convenient to collect the silica on a filter in laboratory test followed by leaching with HF and digestion with boric acid (to remove HF).

Phosphate (ortho: 0.2 to 10 mg/l)

The reagents required for testing is almost as compared to silica testing, except the masking agent. In this case presence of silica needs to be masked, which is done by high concentration of sulphuric acid present in the ammonium molybdate mixture. Testing of this item under TPS practice is relevant for boiler water and cooling water. In the former, phosphate is used as solid alkalisation of boiler water, and in the later organo phosphate/complex forms of inorganic phosphate or polyphosphates is used to prevent scale and corrosion control in cooling cycle, apart from the fact that orthophosphate (viz., inorganic phosphate) if present in the cooling water may itself be a cause of scale formation. To a 50 ml water sample, 5 ml. of photometre molybdate reagent (prepared by dissolving 28 gms of pure ammon molybdate in 700 ml of water and adding slowly but with constant stirring, 225 ml of 98% sulphuric acid; cooled and made up to 1 litre) is added, mixed well and waited for 5 minutes. Then as in case of silica determination, 2.5 ml of same ANSA mixture is added, mixed well and waited for 10 minutes to develop the stable blue colour of molybdenum blue, the intensity of which is measured at 815 nm wavelength of light and ortho phosphate content is assessed by the precalibrated graph.

For best results, sample should be at 21 to 26^{0}C. Sulphides interfere by forming a blue colour directly with the molybdate and nitrites bleach the blue colour. In water containing high salt levels, low results may be obtained. To eliminate these interferences, the sample needs to be diluted suitably.

Phosphate (ortho + poly):

For determination of this, the condensed form of phosphate (meta, pyro or other polyphosphate) is first converted to orthophosphate by using acid and heat to hydrolyse the sample. For the digestion, 50 ml of suitably diluted sample is taken not to contain more than 10 mg/l as total phosphate. To this 2 ml of 10N sulphuric acid is added, mixed and heated to boiling for 30 minutes. After cooling, it is made up to 50 ml again and then phosphate content is determined as per orthophosphate test above and the results recorded.

Phosphate (ortho + poly + organic):

For this, the same method is adopted as in case of phosphate (ortho+poly) above, but for digestion 1 gm of potassium persulphate is also added, which converts any organic phosphate (generally organo phosphate) also along with all other polyphosphates to orthophosphates, which is determined as above and recorded.

Thus, organo phosphonate = (ortho + poly + organic) – (ortho + poly)

Sulphate:

This is determined on the principal that sulphate ions react in acid medium with barium chloride and form white insoluble barium sulphate turbidity. The amount of turbidity formed is proportional to the sulphate concentration and can be measured by photometre at 450 nm up to 50 mg/l. For concentration more than that, the sample needs to be diluted suitably otherwise determined gravimetrically, which is not in practice at a TPS.

50 ml of filtered sample is taken in a 250 ml conical flask and neutralised with 2N HCl and put 2 ml in excess and heat the same to 50 - 60 ^{0}C. To this, while the sample is still hot, 5 ml of barium chloride solution (10%) is added in fine stream with a pipette with constant stirring. A fine white precipitate formed makes the water sample turbid, and the turbidity is measured by absorbance at 450 nm and the sulphate in mg/l is read off a calibration curve of absorbance verses concentration.

Precaution is however necessary to be taken to keep the precipitate slightly warm with agitation, which will not allow the fine precipitate to coagulate and settle.

Nitrate:

Nitrate ion is present in natural water in relatively small quantities, but its estimation is important in view of the fact that any nitrogen compounds introduced by sewage in the water is subsequently converted to nitrate by bacteriological oxidation. Thus, water, which is poor in ammonia & rich in nitrate indicates water has been effectively filtered and purified in the soil.

But high nitrate concentration above 10 mg/l, if used for drinking water, may cause in infants a disease called 'methemoglobinemia' (blue babies).

The basic reaction between nitrate & 2, 4-phenol disulphonic acid produces 6-nitro-2, 4-phenoldisulphonic acid, which upon conversion to the alkaline salt, yields the yellow colour employed for the photometric estimation at wave length of 410 nm. Chloride must however be reduced beforehand to 10 mg/l.

To a 50 ml of water sample, 3 ml. of acetic acid 3 g urea is added and mixed & boiled for 15 minutes. Cooled & made up to 50 ml. This becomes the nitrite free sample. To this 0.5 g of $AgSO_4$ is added to precipitate chloride. Then to it 1 ml of 25% w/v phenol disulphonic acid is added and swirled to wet all deposits. It is then allowed to stand for 10 minutes and washed into a nessler tube. 10 ml of 10% ammonia is then added to it and made up to 50 ml and absorbance is then measured using 410 nm light wave length.

Nitrite:

Nitrite is the intermediate conversion products of ammonical nitrogen by bacteriological oxidation before finally being converted to nitrate and thus its presence indicates existence of contamination of decomposed products of organic matter/ ammonical nitrogen.

To a 50 ml water sample 3 ml acetic acid and 5 ml 20 vol. H_2O_2 is added, mixed and boiled for 15 minutes. This oxidises all the nitrite as present to nitrate. This is then made up to 50 ml and estimate for total nitrate as per procedure described above for 'nitrate'. The difference gives the amount of nitrite present in the sample.

Chloride (in boiler water):

To a 10 ml sample, 0.05 ml 30% v/v perchloric acid and 1.5 ml of 0.5% (in methanol) $Hg(SCN)_2$ (mercuric thiocyanate) are added and mixed. Then 2.0 ml of ferric perchlorate solution (prepared by mixing 300 ml of 60% perchloric acid. 200 ml water and 180 gm of pure ferric perchlorate and made up to 1 litre) is added and mixed. Then this treated sample is kept in dark for 20 minutes and the absorbance at 460 nm wave length using 10 mm cuvette cell is measured. The colour complex is of $Fe\ (SCN)^{2+}$.

NOTES:

1. Do not allow the sunlight to fall on treated sample.
2. Range of application 0 - 10 mg/l.

Hydrazine (low level):

Hydrazine usually in ppb level in boiler feed water is determined on the basis of the reaction of hydrazine in acid medium with p-dimethy amino benzaldehyde to form a yellow colour of azine complex, which is proportional to the hydrazine concentration, the absorption of which is measured at 455 nm. The range of application is 0 - 0.1 ppm N_2H_4.

The sample of water containing hydrazine is collected in acid medium usually HCl to arrest the hydrazine in hydrazine chloride form, so that it does not get vanished by reacting with atmospheric oxygen. 40 ml of such sample is taken for test in a conical flask in such a way that strength of whole solution is 0.01N HCl at 30^0C. To this 10 ml of 3% p-dimethyl benzaldehyde (dissolved in 2N HCl) is added and mixed and intensity of yellow colour is measured after 15 minutes at 455 nm wavelength using 40 mm cuvette cell (light path length) and amount of hydrazine is assessed with previously calibrated graph using same methods of test including for blank.

The standard hydrazine solution is prepared from pure hydrazine sulphate (0.203 g dissolved in 1 litre water including 20 ml of 5N HCl also in it).

Ammonia:

With alkaline nesseler reagent, ammonia gives yellow colour characteristics of low ammonical nitrogen (0.1 to 5 ppm) and reddish brown hues typical of ammonia nitrogen levels approaching 10 ppm. The intensity of colour is then measured at 400 to 425 nm (for yellow colour) and 450 to 500 nm (for brown) wave length using 40 mm light path in the photometre.

Nessler reagent is prepared either with mercuric chloride or with mercuric iodide in the following way:

1. Dissolve 35 g of KI in 100 ml of water. To it add 4% $HgCl_2$ solution with constant stirring until slight red precipitate is obtained (about 325 ml is required). Add to this mixture, a cold freshly prepared solution of NaOH (full amount of 120 gms dissolved in 250 ml). Make-up the whole volume to 1 litre.

   Then add to it little more $HgCl_2$ till a permanent turbidity occurs and stand for 24 hours. When all precipitate is well settled, the clean liquid is slowly decanted and stored in a dark coloured bottled and can be used atleast for one month.

2. Solution A > 160 g NaOH in 500 ml of water & cool. Solution B > 100 g HgCI + 70 g KI dissolved in 250 ml of water. Add solution B to solution A slowly with constant stirring and make the volume to 1 litre and store.

In case of water sample as taken from boiler feed - steam cycle (sample collection must be at less than 40 ^{0}C and ammonia determined immediately), direct nesselerisation is made immediately after collection by adding 1 ml of nesseler reagent to 100 ml of water sample and the absorption noted after due correction for blank reading and the concentration of ammonia read from the precalibrated graph in the same way from the different standards.

However, for other water including drinking one, which produces a turbity or precipitate with nessler reagent, it is better to add 1 drop of 0.05 ml. EDTA reagent (50 g EDTA + 10 g NaOH per 100 ml solution) or 1 to 2 drops of rochelle salt solution (50 gms potassium sodium tartarate tetrahydrate per 100 ml) to prevent precipitation of calcium and magnesium ions in presence of alkali. Then mix well and add 2 ml of nessler reagent to 100 ml of sample, if EDTA has been used.

Alternatively, in boiler feed-steam cycle water, ammonia is also determined by treating the sample with acetone, sodium phenate and sodium hypochlorite, and the concentration of the blue indophenol dye produced is measured absorptiometrically. EDTA is also added to prevent interferences by metal ions including iron. Range of application of this method is 0 - 0.5 mg/l. For this, to a 25 ml sample in stoppered bottle, add 1 ml of 6% EDTA and 0.3 ml acetone. Mix well and add to it 10 ml. freshly prepared sodium phenate (62.5 g phenol + 133 ml 5N NaOH) and 5 ml. of 1% available solution immediately. Stopper the bottle & place at 25 ^{0}C bath and measure the absorbance at 630 nm.

Iron (in boiler feed water-steam cycle):

Iron exists in two states in solution - the ferrous and ferric forms. The method of test described here, though specific to determine Fe (ous) form, but is utilised for determination of all soluble and precipitated forms of iron, viz., total iron after due conversion of precipitated form into soluble by acid and then all Fe (ic) to Fe (ous) by hydroxyl ammonium chloride. Then reaction of iron (ii) with 1, 10 phenonthroline gives an orange red complex $[(C_{12}H_7N_2)_3 Fe]^{2+}$ in proportion to the iron concentration.

The best & stable intensity of colour is at pH 3.5+/- 0.1 and maximum absorption of the colour at 534 nm. The method is applicable over the range of 0 - 50 ppb with sensitivity 10 ppb Fe equivalent to 0.12 absorbance.

Interferences are with Cu, as 10 ppb equivalent to 0.5 ppb Fe. None of other substances likely to be present in steam cycle cause significant interferences. However, the yellow Fe (iii) complex, if Fe (iii) is not reduced properly, will interfere having absorbance at 396 nm. Rather, at 396 nm, both the iron (ii) reddish orange complex and the yellow iron (iii) complex have identical absorption, the amount being additive. The solution slightly acid with sulphuric acid, is treated with 1, 10 - phenonthroline and buffered with potassium hydrogen phthalate at a pH of 3.9; the reading at 396 nm gives the total iron and that at 534 nm the iron (iii).

50 ml of sample (the original collected in a 500 ml polythene bottle already having 10 ml. of 1:1 HCl and the sample collected without overflowing the bottle) is taken in a 100 ml. volumetric flask. To it, 1 ml of 1:1 HCl and 2 ml of hydroxylamine hydrochloride solution (10%; pH of which adjusted to 3.5 either by adding 1:1 ammonia or 1:1 HCl dropwise) is added. The flask is shaken and heated for 1 hour at 60 ^{0}C with its mouth covered with small inverted beaker. After cooling, adjust the pH of content of the flask to 3.5+/- 0.1 with sodium acetate (it is necessary that amount of sodium acetate addition to be ascertained beforehand, so that 0.2 M and 2 M acetate solution is used accordingly without making test aliquot voluminous). To it then 2 ml. of 1, 10 phenonthroline (0.25% in water in monohydrate form) is added and volume

made up to 50 ml and the mixture shaken & allowed to stand for 10 minutes. After that intensity of colour as developed is measured with due correction for blank, the concentration of iron found out with precalibrated graph, for which also the the test procedure had been in exactly the same fashion.

In presence of metallic ions like Cu, Ni, Co, Ag, Bi, etc. are present considerably to make the reddish orange perchlorate complex colour intensity weak, it is better to extract this complex with nitrobenzene and measure the absorbance against a reagent blank.

Iron (in natural water):

The test is based on the determination of iron (iii), which reacts with thiocyanate to give a red coloured complex in true solution, proportionate to Fe (ic) concentration and the intensity of colour is measured at 480 nm absorbance. Fe (ous) does not react. For determination of both Fe (ous & ic) Fe (ous) is first converted to Fe (ic) by oxidising with either hydrogen peroxide or nitric acid. Range of application is generally 20 to 1000 ppb iron. An excess of thiocyanate concentration increases the intensity and also the stability of the colour and comparatively high concentration of acid reduce the influence of presence of phosphate & arsenate. Excess sulphate ions, if present also have a certain tendency to form complexes with iron (iii) ions and hence sulphuric acid use is not recommended under test, nor prsulphate is used for oxidising Fe (ii) ions. The initial boiling with acid removes cyanide & nitrites, which would otherwise interfere.

The sample of water is collected in acid medium in the same way as described in phenonthroline method, to dissolve the precipitate of iron, from which 100 ml of sample is taken in a 250 ml. conical flask with addition of 2 ml. of concentrated nitric acid. The mixture is then boiled till all nitrous oxide fumes are gone off. This oxidises Fe(ous) to Fe(ic). After cooling it make-up to 100 ml volume again and add 5 ml of 4N HCl and 10 ml of 2M potassium thiocyanate. Mix well and stand for 5 minutes and measure the absorbance at 480 nm to find out the concentration of total iron as Fe (iii) from the precalibrated graph made from the different standards exactly in the same way after deducting the blank.

When a large quantity of interfering substances as above are present, it is usually best to proceed in either of the following ways: (1) Remove the iron by precipitation with a slight excess of ammonia solution and dissolve the precipitate in dilute HCl; (2) Extract the iron (iii) Thiocyanate with pure diethyl ether (5:2) and employ the organic layer for the colour comparison.

Total Organic Carbon:

The correlation between ultraviolet (light wave length below 400 nm) absorbance at 254 nm and total organic carbon (TOC) can be determined for a variety of treated and untreated water samples with high correlation coefficients for samples having sufficient organic carbon and lower correlation coefficients with low TOC levels. If interferences from turbidity do not exceed certain limits, even with lower correlation coefficients, accurate TOC determination can be made. However, there is need of judicious selection of light path and wave length of the spectrophotometre, between 210 to 366 nm and a cell length of 40 to 200 mm. The principle is based on the fact that many pollutants commonly found in water strongly absorb ultra violet radiation. It has been generally seen that if the samples are filtered through 0.45 micron millipore membranes, then a close relationship between absorbance at 254 nm using 50 mm light path length and the TOC content is found reasonably good with natural water system. High concentrations of inorganic ultra violet absorbers or high concentrations of organics which do not absorb in the ultraviolet will adversely affect the correlation between the absorbance and TOC. The low pressure mercury lamp has a sharp spectral line at 253.7 nm which, if appropriately filtered, provides a low cost monochromatic light source.

Fluorescence accessory to spectrophotometre works with 190 - 340 nm & with 10 x 10 mm cell, for humic acid & Lignio sulphonic acid.

Organic carbon standard can be prepared with a stable reagent grade organic compound like benzoic acid or anhydrous potassium hydrogen phthalate ($KHC_8H_4O_4$). Usually for the purpose potassium hydrogen phthalate (KHP) is used by dissolving 4.25 gms of KHP in 1 litre of CO free water in a volumetric flask, the solution giving 2000 mg/l organic carbon standard, from which several standards can be prepared by dilution.

Copper (in boiler feed-steam cycle):

The sample is collected in a polythene bottle already having acid just like sample for iron testing (10 ml 1:1 HCl in 500 ml bottle without overflowing it) and kept as it was for 24 hours to allow insoluble forms of copper to dissolve. Then copper is extracted as its dibenzyl-dithiocarbamate complex into carbon tetrachloride, and its concentration determined absorptiometrically at 438 nm. These measures all cupric, cuprous and those other forms of insoluble copper and tested over range of 0 to 50 ppb level with sensivity 10 ppb of copper equivalent to 0.14 absorbance.

The interferences are caused by cobalt, molybdenum and tin (100 ppb giving an effect of 1 ppb additional copper).

A suitable aliquot from acidified sample (200 ml) is taken in a separating funnel, to which 15 ml of 0.1% w/v zinc dibenzyl-dithiocarbamate (DBC) in carbon tetrachloride is added. Shake the funnel vigourously for 1 minute and set aside for atleast 10 minutes to allow the two phases to separate. The carbon tetrachloride layer is then taken out and absorbance measured at 438 nm using 40 mm cuvette cell within 60 minutes and do not allow direct sunlight to fall on the extracts. The copper content is then known from the precalibrated graph having necessary correction of blank carried out in the same way.

Alternatively, use of neo-cuproine (2, 9 dimethyl 1, 10 phenanthroline) is also made, which reacts with copper and produce yellow coloured cuprous complex, the intensity of colour being proportional to the copper content. Full development of colour takes place over the pH range from 2.3 to 9 and the range of concentration measured is from 2 to 1000 ppb.

200 ml of acidified sample is taken in a 250 ml separating funnel and to this 1 ml hydroxylamine hydrochloride solution (40 g per 200 ml water and copper present, if any, removed by treating it in a separating funnel with neocuproine and chloroform solvent) is added and mixed by shaking. To it then 10 ml of sodium acetate solution (275 gms per litre of water and duly removed from copper) is added and mixed. After that 2 to 4 ml of neo-cuproine solution (by dissolving 0.1 g in 50 ml of isopropyl alcohol & diluted to 100 ml with water) is added and again mixed. To this, then 20 ml. of chloroform solvent (9 volumes of chloroform mixed with 1 volume of isopropyl alcohol) is added and the funnel is shaken vigorously for 1 minute,

and is allowed to stand for 5 minutes. The chloroform layer is then drained into a 50 ml Erlenmeyer flask, clearing the solution by further addition of 5 ml of isopropyl alcohol. The absorbance of this extract is then measured at 457 nm and compared with precalibrated graph under identical procedures.

In place of chloroform solvent, iso amyl alcohol can be used. In this case however drain the aqueous layer & discard.

Instrumental Methods

Flame Photometry Sodium & Potassium:

In the temperature range of 1600 - 1700 ^{0}C (as could be obtained by burning domestic LPG gas associated with compressed air) certain metallic ions like of sodium, potassium, lithium and calcium are excited, when their solution (particulary in chloride form) is sprayed as a fine mist into a non-luminous flame resulting in raising of an electron to a higher energy level (electron is obtained by decomposition of dissociated atoms into flame) followed by the loss of a discrete amount of energy in the form of radiant energy as the electron falls back into its original position or at a lower energy level. In the process, it emits light of characteristic wave length, which when isolated to the characteristic wave band by an optical filter, as below, and allowed to fall on a photo cell, it gives specific electronic signal, which is a measure of concentration of the element and is measured by a suitable detector

e.g. electronic amplifier and a metre / galvanometre:

Sodium, potassium, Lithium, calcium	Yellow doublet at 589 nm red doublet at 767/769 nm single lines at 671 nm this is not excited in the above flame, as it requires higher energy. However red band of Ca at 626 nm is used for determination of calcium by flame photometry

Before analysing the unknown solutions, the system is standardised with solutions of known concentrations.

In the instrument, due to a draught of air at the tip of the atomiser, the sample solution in the form of chloride is sucked in and enters in the mixing chamber of the instrument as a fine atomised jet. The mixture of gas and atomised sample is passed on to the burner and is ignited. The emitted light from the flame is collected by a lens and is passed through an appropriate filter as above. The filtered light is then passed on to energise a sensitive photodetector, the output of which is applied to the electronic metreing unit for read out. If the signal is found weak, use of an additional accessory called spot galvanometre may be made, which will amplify the signal suitably to measure sodium and potassium in ppb level also.

Instrumental Methods

(Selective Ion Electrodes)

Sodium (in Boiler Feed—Steam cycle):

A specific ion electrode, viz., sodium electrode when immersed in a sample of water along with a reference electrode, forms an electrochemical cell whose potential is dependent on the sodium ion activity in the sample, the cell responding to the following modified Nernst equation.

Thus, when these electrodes are coupled to a pH metre having concentration scale (p Ion metre) gives direct concentration read out of sodium in the sample in quick time and eliminates the calibration graph also, the calibration is only necessary once a day using one standard solution. Also, the electrode requires little or no maintenance and if stored correctly has a very long working life.

For the purpose of boiler feed-steam cycle water sample testing, specific low level sodium electrode are employed with refillable calomel type reference electrode, which can generally measure sodium content down to 0.1 ppb level.

Significant interference however occurs from hydrogen ions if $pH < pNa + 3$ and from silver ions if $Ag^+ > 10^{-4} \times Na^+$. Also, since response of the electrodes is temperature sensitive, samples and standards must be of similar temperature. In view of silver ions are not encountered in thermal power station chemistry, the interference due to hydrogen ions is removed by treating the sample with an alkaline buffer such as ammonia buffer. For optimum response the electrode should also be used in a continuous flow system and contact with glass apparatus should be avoided.

Chloride (in boiler feed-steam cycle):

Low level chloride is also measured by selective ion electrode, viz., Ag/AgCl, the potential of which when immersed in the sample, is measured and the concentration of chloride is obtained from the calibration graph. Range of application is 0.1 to 10 ppm chloride and almost linear over this range.

The chloride electrode along with the reference electrode are suitably clamped to put in a 50 ml beaker having polythene covered magnet piece (20 mm long) at the bottom and kept on the platform of varying speed suitable magnetic stirrer to rotate the magnetic piece slowly and steadily in the beaker. Then the electrode along with stirring rod is washed suitably into the beaker with three small portions of same solution, whose potential is to be measured. After throwing the rinsed liquid, the surplus solution from the reference electrode is dried out by touching lightly with a soft tissue. Lift the electrodes above the beaker level and connect the same to a milli voltmetre. Then a standard chloride solution (say 10 ppm) equilibrated at a water bath temperature (room temp (+/- 2 ^{0}C) in sufficient quantity and already added with 0.5 M acetic acid/0.5 M ammonium acetate buffer solution (buffer: solution = 1:9) is poured in the beaker so that both electrodes are fully immersed, and the solution stirred. The bottom of the reference electrode should be slightly lower than that of the chloride sensitive electrode to safeguard the later against damage. The electrodes are then put in the solution and e.m.f reading is noted after 2 minutes and recorded. Remove the electrodes and dry the surplus solution from the reference electrode and empty the beaker to clean.

Repeat the above procedures with 1 ppm standard chloride solution and then with the sample solution exactly in the same ways separately. Calculate first the mean e.m.f's of the standard chloride solutions, and plot these values against (log ppb Cl + 1) & draw a line through the plotted means of the standard solutions. Then read off these values (of ppb Cl +1) corresponding to the measured emf of the samples and hence obtain their concentrations.

Since the chloride concentrations being measured may be below the solubility of silver chloride, the electrode will need re-chloridising periodically. The silver chloride surface of the electrode is washed with dilute ammonia until the matt silver surface reappears. Then immerse the silver electrode along with platinum electrode in 1:99 HCl and electrolysis conducted employing

0.2 mA current. After electolysis, remove the electrode from the cell and wash thoroughly with pure water for 1 - 2 hours and store in the pure water in dark when not in use. It is recommended that electrodes are not used for the first week after chloridising.

Copper (in boiler feed-steam cycle):

Low level copper is also measured by selective ion electrode, viz., cupric electrode, which consists of a membrane containing sulphides of copper and other metals bonded in an epoxy body. When the membrane is in contact with a cupric solution, an electrode potential develops across the membrane. This potential which depends against a constant reference potential (liquid junction) with a specific ion metre, mv metre, pH metre. The measured potential correspond to the level of cupric ions in solution, obeying Nernst equation:

E	=	E0 + S log (A)
where E	=	Measured electrode potential
E0	=	Reference potential (a constant)
A	=	Cupric ion level in solution
S	=	Electrode slope (about 28 mv per decade)

In neutral solution, cupric ion concentration can be measured by this electrode down to 0.6 ppb level mercury and silver ion poison the electrode sensing element, but these are generally absent in normal operation of boilers in TPS. Ferric ions affect the membrane surface, only if the ferric ion level is greater than one tenth the cupric ion level. However, ferric ion can be eliminated from the sample simply by adding sodium fluoride.

The sample water must be below pH 6 to avoid precipitation of $Cu(OH)_2$ and hence is needed to be collected in dilute nitric acid of suitable strength and pH further adjusted with 1M NaOH/1M HNO as the case may be. Further, 2 ml of $NaNO_2$ (5M) per 100 ml of sample water is also added as ionic strength adjuster (ISA) and potential measured.

The concentration is then calculated out by the plotted graph made by the standard solutions corrected to blank and potential measured exactly in same way. The temperature of the sample and standards should also be the same. A 1 ^{0}C difference in temperature will give rise to about a 4% error.

Instrumental Methods

Filtration technique Iron or crud:

Undissolved iron in the form of crud is measured by membrane filtration utilizing 0.45 micron porosity, 47 millimetre diametre millipore membrane. 1 to 4 litres are filtered through this and the stain on the millipore membrane is compared to the B & W membrane filter comparison charts. No stains mean iron is below 5 ppb.

Interpretation of Analytical Data

On water analysis

Any collection of analytical data is meaningless unless an informed and logical interpretation can be made. Keeping in view of this, the following could be noted:

Checking correctness of analysis:

1. Anion-Cation Balance:

   Theoretically, the sum of the anions (as me/l) must equal exactly the sum of cations (as me/l), but in practice these are not equal. The unequality increases as the ionic concentration increases. However, for normal surface water, the difference is generally not more than 0.1. For other water, the acceptable units are generally given by:

   Total Anions (each ion measured as me/l) - Total Cations (each ion measured as me/l) = +/- (0.1065 + 0.0155 of Total anions)

   'me/l' (milli equivalent per litre) or 'epm' (equivalent per million) are ppm (parts per million) units divided by the equivalent weight of the ion, and is a measure of concentration in terms of its power to react with other ions.

   Since, some of the constituents of water like total dissolved salts total suspended matters, organic matters, gases, etc. must be reported in ppm only, parts per million remains as common standard way of reporting a water analysis except when extensive calculations are required. The main drawback to the epm unit is that in a water analysis of normal accuracy, it runs to one or more decimal places, which makes it a little unwieldy, and thus expressing results in ppm remains popular.

Hence, a more common way to measure equivalent concentration is to express the results in 'ppm as $CaCO_3$', in which every ion is calculated as the chemically equivalent concentration if it were $CaCO_3$. Thus, the same water analysis if now calculated by taking each of the constituent salts in ppm as such, dividing by its equivalent weight and multiplying by 50 (equivalent wt of $CaCO_3$), then it can be easily seen that Total cations ppm as CaCO3 becomes nearly equal to the total cations ppm as $CaCO_3$ giving exactly the same information as given by the epm unit.

As a matter of fact, this is merely the epm analysis multiplied by 50, and we are dealing in whole numbers rather than decimals. Thus, we are using both systems of expression of results (epm or ppm) within one analysis.

NOTES:

1. One should not be confused with the concept that while $CaCO_3$ is splendidly handy in having an equivalent weight of exactly 50, it is not actually soluble in that form and in view of this, there could be error in measuring bicarbonate & carbonates 'as calcium carbonate'. It should be remembered that that 'as calcium carbonate' bit is just a convention, and in practice is simply a 50 fold multiplier.

   Under 'normality' system, viz., me/l is a weight/volume system, by contrast to the epm system, which is weight/weight. At N concentration, the density of solutions is significantly greater than 1.00, and therefore N and me/l are not strictly interchangeable with epm. In practice, the error is a lot smaller than the percentage error inherent in most water analysis, and for normal purposes the distinction can safely be ignored.

2. **Total Dissolved Salts:** To calculate dissolved solids, convert the various constituents from ppm to epm and total the various cations (positively charged ions) and anions (negative ions). The cation should equal the anions. If not, add either sodium (plus) or nitrate

(minus) ions to balance the columns. Then convert each component ionic epm to ppm and total to obtain ppm dissolved salts. However, for practical purposes, when total cations and total anions both expressed ppm as $CaCO_3$ are found approximately equal, then the total anions summed with total silica is taken as total dissolved salts.

3. **Conductivity:** The total dissolved salts are also related with electrical conductivity (which is an approximate measurement of the total ionised salts/impurities present in water. Usually, conductivity (micromhos per cm or microsiemens per cm at 25 ^{0}C) multiplied with 0.55 to 0.7 is found equal to mg/l total filterable residue. For water, that contain appreciable concentration of free acid or caustic alkalinity, the factor may be much lower than 0.55 and for highly saline water, it may be much higher than 0.7. This, however, usually may be roughly applied to water having pH between 6 and 9.

   The following are the conventional conductivity factors of ions commonly found in water:

Ions	Conductivity (me/L)	Microsiemens/cm (25 ^{0}C) per mg/l
Bicarbonate	43.6	0.715
Calcium	52.0	2.60
Carbonate	84.6	2.82
Chloride	75.9	2.14
Magnesium	46.6	3.82
Nitrate	71.0	1.15
Potassium	72.0	1.84
Sodium	48.9	2.13
Sulphate	73.9	1.54
1 ppm of CO_2 contributes about 0.05 micromhos/ cm conductivity.		

4. **Total Cations by Ion exchange:** Suitable aliquot containing 0.1 to 0.2 me/l cations made up to 100 ml volume & taken in a 250 ml

conical flask. Add 2 g of cation exchange resin in H^+ form and stir at moderate speed for 15 minutes. Filter the resin through glass wool & wash with two portions of 15 ml water. Titrate the combined filtrate & washings to pH 4.5 with 0.02 N NaOH.

Determine with fresh aliquot the alkalinity in the sample in me/l by titrating to pH 4.5 with 0.02 N standard acid.

5. **Alkalinity & Non-carbonate Hardness:** These individual to respective ions calcium, magnesium & sodium can be found as below:

Calcium alkalinity	Ca hardness or alkalinity, whichever is smaller (if equal, either one)
Magnesium alkalinity	Mg hardness, if alkalinity is equal to or greater than total hardnes
	Alkalinity - Ca hardness, if alkalinity is less than total hardness
Sodium alkalinity	Alkalinity - total hardness
Ca non - carbonate hardness	Ca hardness - Ca alkalinity
Mg non - carbonate hardness	Mg hardness - Mg alkalinity
Total non - carbonate hardness	Total hardness-alkalinity

NOTES:

1. In the above, if any computation yields a zero or negative result, none of that substance is present.
2. Bicarbonates & hydroxides do not exist together nor does carbon dioxide with carbonates. 1 ppm as $CaCO_3$ of bicarbonate alkalinity produces 2 ppm of CO_2 as $CaCO_3$. Also, carbonate cannot be present at pH below 8.2.
3. If pH < 3, means less probability of presence of carbonate, and if pH < 4.5, means presence of both carbonates & bicarbonates are doubtful.
4. Soluble iron & manganese are generally present in bicarbonates form. Hence, bicarbonate alkalinity should be accounted for these two items also.

5. For total hardness Ca:Mg ratio in normal surface water is generally assumed to be in the range 60:40. Hardness is also contributed by Fe, Al, Zn and Mn and can be expressed as mg/l as $CaCO_3$ by multiplying its respective metallic contents by 2.69, 5.567, 1.531 and 1.822 respectively.

6. HCO_3 ions ppm x 0.82 + CO ions ppm x 1.66 = total alkalinity ppm as $CaCO_3$

7. Calculate all the calcium, up to the limit of the alkalinity as $Ca(HCO_3)_2$ Then calculate any remaining calcium as sulphate, nitrate, nitrite and finally chloride in this order. Any alkalinity remaining should be calculated as $Mg(HCO_3)_2$ and the remaining magnesium as sulphate, nitrate, nitrite and finally chloride in turn. The silica should be calculated as sodium silicate. If any alkalinity is unaccounted for, the difference should be recorded as sodium bicarbonate. Any remaining sodium should then be calculated as sodium chloride.

8. **Others**

   a) About 1 ppm of dissolved oxygen will oxidise about 7 ppm ferrous iron & manganous manganese. Ferrous and manganous ions are generally present in tubewell water, which are virtually free from dissolved oxygen. These are also found in polluted water.

   b) If free ammonia is present, one must look for albuminoid ammonia, from which ammonia can further be released under specified conditions from the organic matter (nitrogeneous) still existing in the sample.

   c) When chloride content is fairly high, it is usually associated with Na as common salt. When river water is free from any other major source of chloride, the chloride content is the simplest chemical indication of the amount of sewage effluent, which has been added.

   d) The amount of CO_2 in water in equilibrium with the atmosphere is about 0.5 mg/l whereas amounts up to 80 mg/l are found in some water from underground sources. For the the tentative relation between free CO_2 & pH at equilibrium in the form of 'table' may be referred to 'cooling water chemistry' chapter (under the head 'scaling'). Similarly, water at a temperature of 20 ^{0}C and a pressure of 1 bar will have 9.1 ppm of dissolved oxygen at equilibrium.

e) For normal alkalinity, pH & Na content, true soluble SiO_2 should not be more than 20 mg/l. If higher, means some colloidal silica should also be present.

f) Taste and odour of water are generally due to the following and any investigation should be made accordingly:

Decaying vegetation e.g. algae, etc. Moulds & Actinomycetes	Grassy, fishy or musty odours Earthy, musty or mouldy taste
(Have favourable condition of growth long length)	(1st water drawn in warm surroundings in morning will give pipe lines of large buildings) very unpleasant odour
Iron & sulphur bacteria decomposition	Offensive smell, sometimes like rotten eggs
Excess NaCl Iron Industrial wastes Chlorine	Brackish bitter taste characeristics smell do

g) $KMnO_4$ is not the volume or weight of organic matter present, but the amount of oxygen absorbed in the test. This indicates the amount of carbonaceous organic matter (and not nitrogeneous) in a water, and is purely empirical one for estimating the amount of organic matter present.

9. **Miscellaneous**

a) On hardness basis, quality of water can be graded as below:

Total hardness of water, ppm as $CaCO_3$	Nature
0 to 50 ppm	Soft
50 to 100 ppm	Moderately soft
100 to 150 ppm	Slightly hard
150 to 200 ppm	Moderately hard
> 200 ppm	Hard
> 300 ppm	Very hard

NOTE: Very soft water must be examined for plumbo solvency test, in view of its high corrosiveness.

b) On the following grounds, water quality gradation could be as below:

Calcium alkalinity source	Excellent source	Good source	Poor source	Rejectable source
Average BOD 5 days, mg/l	0.75 - 1.5	1.5 - 2.5	2.5 - 4	> 4
Max BOD in any one sample, mg/l	1 - 3	3 - 4	4 – 6	> 6
Average coliform,	< 5% of	< 10% of	< 5% of	--
MPN/100ml pH	samples	samples	samples	
	6 - 8.5	5 - 6	3.8 - 5	< 3.8
		8.5 - 9	9 - 10.3	> 10.3
Chlorides mg/l	< 50	50 - 250	250 - 600	> 600
Fluorides	< 1.5	1.5 - 3	> 3	--

c) Soluble inorganic compounds commonly encountered in industrial water supplies are as below:

i. Bicarbonates of Ca, Mg, Na, Fe (ous) & Mn (ous)

ii. Chlorides of Ca, Mg & Na

iii. Sulphates of Ca, Mg & Na

iv. Silica in the forms of SiO_2, Na SiO & H_2SiO_3

NOTES: In ion exchange, although the silica in water is not ionised, it is removed by formation of the silicate form of the quaternary ammonium anion exchange resin, the 'adsorbed' species being monovalent H_2SiO_3 (refer demineralisation section)

S I PREFIXES

Prefix	Symbol	Multiplication factor
Exa	E	10^{18}
Peta	P	10^{15}
Tera	T	10^{12}
giga	G	10^{9}
mega	M	10^{6}
Kilo	k	10^{3}
hector	h	10^{2}
deca	da	10
deci	d	10^{-1}
centi	c	10^{-2}
milli	m	10^{-3}
micro	u	10^{-6}
nano	n	10^{-9}
pico	p	10^{-12}
femto	f	10^{-15}
Atto	a	10^{-16}

Calcium Carbonate Equivalents

Conversion factor

Ion	Equivalent weight	Substance to $CaCO_3$ equivalent	$CaCO_3$ equivalent to substance
Cations			
Al^{3+}	9.0	5.56	0.18
NH_4^{+}	18.0	2.78	0.36
Ca^{2+}	20.0	2.50	0.4
Fe^{2+}	27.9	1.79	0.56
Fe^{3+}	18.6	2.69	0.37

Ion	Equivalent weight	Substance to $CaCO_3$ equivalent	$CaCO_3$ equivalent to substance
H^+	1.0	50.0	0.02
Mg^{2+}	12.2	4.1	0.24
Mn^{2+}	27.5	1.82	0.55
K^+	39.1	1.28	0.78
Na^+	23.0	2.18	0.46
Zn^{2+}	32.7	1.53	0.65
Anions			
HCO_3 -	61.0	0.82	1.22
CO_3^{2-}	30.0	1.66*	0.60
Cl^-	35.5	1.41	0.71
NO_3-	62.0	0.81	1.24
OH^-	17.0	2.94	0.34
PO_4^{2-}	31.7	1.58	0.63
SiO_2	30.0#	0.83	0.60
SO_4^{2-}	48.0	1.04	0.96

*For ion exchange calculations, it is assumed that carbonate exchanges as the monovalent ion and the conversion factor is 0.83.

#Silica in the form of HSiO is exchanged by anion resin, which is monovalent and hence its equivalent weight is taken as 60 for all calculation purpose.

www.ingramcontent.com/pod-product-compliance
Ingram Content Group UK Ltd.
Pitfield, Milton Keynes, MK11 3LW, UK
UKHW041623031026
15587UKWH00001B/20

9 789355 276773